ライフサイエンス

文例で身につける

英単語・熟語

著／河本　健, 大武　博
監修／ライフサイエンス辞書プロジェクト
英文校閲・ナレーター／Dan Savage

羊土社
YODOSHA

【注意事項】本書の情報について

本書に記載されている内容は，発行時点における最新の情報に基づき，正確を期するよう，執筆者，監修・編者ならびに出版社はそれぞれ最善の努力を払っております．しかし科学・医学・医療の進歩により，定義や概念，技術の操作方法や診療の方針が変更となり，本書をご使用になる時点においては記載された内容が正確かつ完全ではなくなる場合がございます．また，本書に記載されている企業名や商品名，URL等の情報が予告なく変更される場合もございますのでご了承ください．

まえがき

近年のTOEICブームで英語学習熱はますます高まっている．しかし，TOEICで高得点することよりも，自分に必要な英語力を身につけることの方がもっと重要ではないだろうか．著者らは，ライフサイエンス分野の専門英語の参考書として，「ライフサイエンス英語類語使い分け辞典」「ライフサイエンス英語表現使い分け辞典」「ライフサイエンス論文作成のための英文法」を製作した．これらは主に論文執筆時などに参照するものだが，そうは言っても，その時だけ安直に利用して何とかなるというほど甘くもないのも事実である．そこでもう少し気軽に専門英語の勉強に取り組むための学習書が必要性だと考えて本書を作成した．医学・生物学の分野の教科書や論文を読んだり，学会抄録や論文を書いたりする英語力を身につけるためには，それに基づいた学習書を利用することが望ましいからだ．

英語のリーディングやリスニング時に，一番ネックになるのが単語力であることはほぼ間違いないだろう．キーワードが理解できなければ，結局は何のことを言っているのかわからない．もちろん辞書を引けばそれで済むことも多いが，いちいち辞書を引いていては能率が上がらないし，英語力も向上しにくい．また，文法的な知識が不足していれば，ひとつの文に含まれる単語の数が多い論文の文章を正しく理解できない．このように単語力や英文法力の増強は，緊急の課題であることに間違いない．

本書では，ライフサイエンス分野の論文などでよく使われる動詞・形容詞・副詞・名詞の用法の中で特にマークすべき重要なもの，あるいは特徴的なつなぎ表現・比較表現・熟語表現などをピックアップして分類し，それを使った415例文を収録した．例文には，生命科学分野の論文や教科書でよく使われる英単語1,462，連語表現・熟語・複合語795が含まれている．本書は，従来の英単熟語の学習書とは大きく異なって，語法を中心とした構成に対する例文を単語学習用の例文と兼ねる構成になっている．これによって単に単語の意味を覚えるだけでなく，よく使われる構文や表現方法をまとめて学習できるメリットが生まれてくる．前著の内容とも関連づけた構成になっているので，合わせて利用していただければ学習効果もさらに上がるであろう．

本書の特徴の1つとして，すべての例文に対する音声教材が羊土社のホームページからダウンロードできることがある．これを活用して，音声を中心とした繰り返し学習を行えば非常に効果的であろう．ライフサイエンス分野の専門英語力向上のための本書を役立てていただけることを願っている．

2009年6月

著者を代表して

河本　健

著者 / 河本　健
広島大学大学院医歯薬学総合研究科助教

大武　博
福井県立大学学術教養センター教授

監修 / ライフサイエンス辞書プロジェクト

金子周司
京都大学大学院薬学研究科教授

鵜川義弘
宮城教育大学環境教育実践研究センター教授

大武　博
福井県立大学学術教養センター教授

河本　健
広島大学大学院医歯薬学総合研究科助教

竹内浩昭
静岡大学理学部生物科学科准教授

竹腰正隆
東海大学医学部基礎医学系分子生命科学講師

藤田信之
製品評価技術基盤機構バイオテクノロジー本部

英文校閲・ナレーター / Dan Savage
コロラド在住，メディアプロデューサ

ライフサイエンス

文例で身につける

英単語・熟語

まえがき
本書の特徴と使い方　河本　健　9
学習のポイント　12
音声教材の活用方法　14
発音を身につけるためには　河本　健　16

I. 動詞の使い方

1 他動詞＋目的語の文例
A. 他動詞＋名詞　[001～024]　20
B. 他動詞＋that節　[025～037]　33
C. 他動詞＋whether節　[038～043]　40
D. 他動詞＋to *do*　[044]　44

2 他動詞（過去分詞）＋前置詞／that節の文例
A. 他動詞（過去分詞）＋前置詞＋名詞　[045～060]　45
B. 他動詞（過去分詞）＋to *do*　[061～071]　54
C. itを形式主語とするthat節　[072～073]　59

3 自動詞＋前置詞の文例
A. 自動詞＋前置詞　[074～095]　60

II. 副詞の使い方

1 副詞＋過去分詞の文例　[096～119]　71
2 副詞＋形容詞の文例　[120～131]　84
3 副詞＋前置詞の文例　[132～146]　90
4 文頭の副詞の文例　[147～155]　99

III. 形容詞の使い方

■ 形容詞＋前置詞の文例　[156～180]　104

IV. 名詞の使い方

1 名詞＋前置詞の文例 [181～234] 117

2 名詞＋that（同格のthat）の文例 [235～238] 145

V. つなぎの表現

1 逆説の文例

A. 副詞／副詞的熟語 [239～247] 148

B. 副詞節を導く接続詞 [248～251] 153

C. 副詞句を導く熟語／接続詞 [252～257] 156

2 肯定の文例

A. 副詞／副詞的熟語 [258～271] 160

B. 副詞節を導く接続詞／熟語 [272～274] 168

C. 副詞句を導く熟語 [275～282] 170

3 まとめの文例

A. 副詞／副詞的熟語 [283～285] 174

4 条件の文例

A. 条件 [286～289] 176

VI. その他の表現

1 比較の表現の文例

A. ～ than [290～296] 179

B. ～-fold, %, times [297～304] 183

C. compared, comparison, relative [305～307] 187

D. to a lesser ～／… degree of ～ [308～309] 189

2 asを用いた表現の文例

A. as＋過去分詞 [310～326] 191

B. as ＋名詞 [327～345] 201

C. as ～ as [346～356] 211

D. as ＋前置詞 [357～360] 216

E. asで始まる節 [361～362] 218

VII. 熟語表現

■ 熟語表現の文例 [363～415] 220

本書に掲載の文例一覧 246

索　引 275

音声教材ダウンロードのご案内

本書に掲載している文例を収録した音声教材を
下記手順にてダウンロードできます.ぜひご活用ください

※音声データはご登録いただいた本書の読者のみご利用いただけます

登録・ダウンロード手順

1. パソコンから空メールを eitango@yodosha.co.jp 宛に送信してください
2. 返信メール中の URL をクリックし登録ページにアクセスしてください※
3. 登録画面にて必須事項をご入力ください
4. すべて入力後に送信ボタンを押すと登録内容確認画面が表示されます.内容をご確認のうえ OK ボタンを押してください
5. ご登録いただいたメールアドレスに音声教材ダウンロードページの URL と,ユーザ名・パスワードをご連絡いたします
6. ダウンロードページにアクセスするとダイアログボックスが出ますので,ユーザ名とパスワードを入力してください
7. ページに書かれた案内に従って,必要なファイルをダウンロードしてください

※ご利用の環境によっては"返信メール"が迷惑メールとして処理されるなど正常に受信できない可能性があります.しばらく経っても"返信メール"を受信しない場合,yodosha.co.jp のメールを受信可能にするなど迷惑フィルターの設定見直しを行ってください.設定変更ができない,不明な方は件名を「返信メール未到着」として eitango@yodosha.co.jp までご連絡ください

ダウンロードできるデータについて

- 音声は MP3 形式で作成しております.iTunes, Windows Media Player などの音楽ソフトでご利用ください
- ダウンロードできる音声の録音パターンは下記 2 種があります

【Slow Speed(ゆっくりと読み上げ)】
トラック名[001_1slow, 002_1slow, … 415_1slow]

【Normal Speed(通常のスピードで読み上げ)】
トラック名[001_2normal, 002_2normal, … 415_2normal]

それぞれ別々にダウンロードすることが可能なので,用途に応じて使い分けてください

音声教材の活用のしかたは,"音声教材の活用方法"(p.14) をご覧ください

本書の特徴と使い方

本書は，ライフサイエンス分野の英語の単語や構文の学習書として企画したものである．英語の参考書といえば以前は大学入試用が中心であったが，最近ではTOEICの学習書が溢れんばかりである．しかし，それ以外の目的の英語学習書というと，会話やリスニングなどが中心でかなり限定的と言える状況だ．ましてや，専門分野の英語の学習書となると非常に限られてくる．本書は，ライフサイエンス分野の英語の専門書や論文の読解や執筆に必要な英単語・熟語・構文などの基礎力アップのために役立つものを目指している．普段から気軽に学習して，論文詳読に応用し，さらにいざ，論文執筆というときに備えていただければ幸いである．本書は特別な試験を目的とするものではないが，専門分野の英語の入学試験や資格試験に向けた勉強にも役立つであろう．

本書の特徴

1. ライフサイエンス分野の論文で使われる動詞・形容詞・副詞・名詞の用法の中で特にマークすべき重要なもの，あるいは特徴的なつなぎ表現・比較表現・熟語表現などをピックアップして分類し，それを使った415例文を収録してある．
2. 例文は，ライフサイエンス分野の教科書や論文でよく使われる重要単語1,462，連語表現・熟語・複合語795を用いて作られている．
3. 収録した単語，連語表現，熟語，複合語，重要表現には日本語訳などの解説および6,000万語のLSDコーパス中での出現回数を示してあり，学習の際の目安になる．
4. 収録した単語に関連する重要な類語・反意語が収録してあるので，合わせて習得するとよいだろう．
5. 例文すべての音声教材を羊土社のサーバーからダウンロードして利用できる．

本書の構成

重要な用法一覧

論文でよく使われる動詞・形容詞・副詞・名詞の用法，つなぎの表現，比較表現，熟語表現などを各項目のテーマに沿ってピックアップしてある．

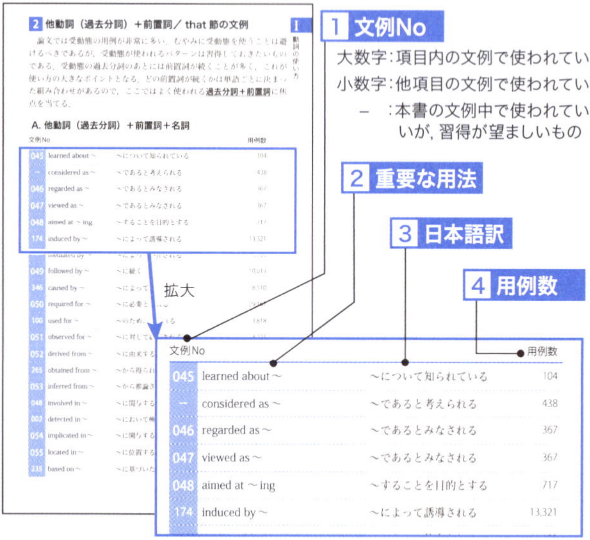

1 文例No	掲載の用法を用いた文例の番号．大きな数字は項目内の重要用法の文例として示されているもの，小さな数字は他項目の文例の中で使われているもの，－は本書の文例内で使われていないが，合わせて習得することが望ましい表現である．
2 重要な用法	論文中でよく使われている重要な用法，表現のパターン．
3 日本語訳	提示した用法の一般的な日本語訳．
4 用例数	6,000万語のLSDコーパスの中での出現回数．どのような表現の使用頻度が高いのかを知ることができる．

文例，単語・熟語

2 で挙げた重要な用法を用いた英文と英文中で使われている単語・熟語をを解説している．

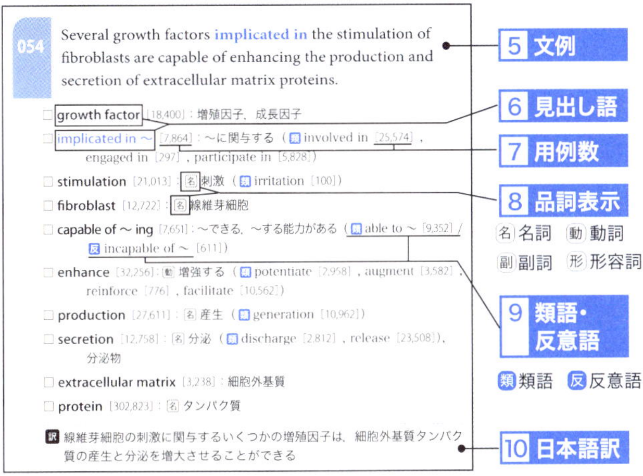

5 文例	論文でよく使われる表現を含む英文．2 で挙げた用法は色文字で示してある．
6 見出し語	文例中に出てきた語のなかで，重要単語やライフサイエンスや医学に特徴的な語を解説している．
7 用例数	見出し語だけでなく，類義語や反意語にも用例数を示してある．
8 品詞表示	単語については名詞・動詞・副詞・形容詞を表記してある．
9 類語・反意語	見出し語の類語・反意語も示してあるので学習に役立てることができる．
10 日本語訳	文例の理解のため，日本語訳を示してある．

学習のポイント

単語力を増強するための方法には多読に勝るものはないとは，誰もが考えることである．たくさん読んでその文脈の流れの中で単語を覚えれば，大学入試用の単語学習書で覚えたものよりも遙かに忘れにくいはずだ．そうはいっても，同じ単語の意味を何度も辞書で調べてしまうことは誰もがしばしば経験することであろう．多読で単語力を増強するためには，相当の量の英文を読まなければならない．また，多読は辞書を引かずに読むことが基本であり，方法論的なジレンマもある．そこで，もっともお勧めできることは再読である．自分の必要な単語がたくさん載っている本を何度も繰り返し読むことだ．広く浅く読んでしまっては，単語力はなかなか付かないが，内容を覚えるぐらい繰り返し本を読めば単語もよく覚えられるであろう．

このように，**単語力増強のための鍵は繰り返し**である．同じ単語の学習を何度も繰り返すことによって初めて効果も上がるのだ．そこで本書は，その単語の繰り返し学習のための教材として作成された．415ある例文は，すべてアメリカ人に読み上げてもらった音声教材があるので，リスニング学習も合わせて行うことができる．slowバージョンで聞いて個々の単語の音を確認し，次にnormalバージョンでリスニングに挑戦するとよいだろう．

単語学習は，とても単調なものである．前後の見出し語との関連もアルファベットの順番や意味の類似性ぐらいしか見いだせない場合が多い．そこで本書では，**論文でよく使われる構文や連語表現に着目し，これらを単語学習と平行して行える**ように編集した．取り上げた項目は「ライフサイエンス論文作成のための英文法」の項目とかなり共通しているので，合わせて学習すると効果的である．収録した例文は必ずしも論文調ではないが，構文はよく使われるものなので，英語論文の読解だけでなく学会抄録や論文執筆に役立つであろう．

本書の収録語（表現）数は，単語が 1,462，連語表現・熟語・複合語が 795 である．さらに例文に登場しない類語・反意語・関連表現が 929 収録してある．合計 3,186 語（表現）あるが，残念ながら膨大な専門用語が存在するなかでは，この語数は十分とはいえないだろう．PubMed 抄録の 90％をカバーする見出し語数 6,000 語の「ライフサイエンス必須英和辞典」も既に発売中であるので，そちらも合わせてご利用いただければ幸いである．

LSD コーパスについて

LSD コーパスは，ライフサイエンス辞書（LSD）プロジェクトが独自に構築したライフサイエンス分野の専門英語のコーパスである．ライフサイエンス分野では PubMed と呼ばれる無料の文献データベースがあるが，そこからライフサイエンスのさまざまな分野を網羅する主要な学術誌（約 100 誌）を選び，1998 年から 2005 年までの間にアメリカまたはイギリスの研究機関から出された論文抄録（総語数約 6,000 万語）を集めてのコーパスを構築してある．LSD コーパスは，LSD プロジェクトのホームページ（http://lsd-project.jp/）から利用できる．なお，本書に収録した例文は LSD コーパスとは無関係である．

音声教材の活用方法

本書に収録したすべての例文には，音声教材が作製してある．羊土社のサーバーから無料でダウンロードできるので，是非，活用しよう（p.8 参照）．ダウンロードできる教材は音声ファイル(MP3 形式)になっているので，iTunes などのパソコンソフトに取り込んで利用できる．さらに CD に焼いたり iPod などの携帯音楽プレイヤーに取り込んで使えば活用する場面が増えるだろう．

学習の方法としては**単なるリスニングよりも，シャドーイングを行う**ことを強くお勧めしたい．シャドーイングとは，教材の音声を少しだけ遅れてリピートすることである．例文の読み上げを聞いてから繰り返したのでは，正確に発音することが困難である．そのため正しい発音を身につけるためにはシャドーイングの方がよい．また，教材の音声のあとの無音部分は２秒程しかないので，全部聞いてから繰り返していては間に合わないかもしれない．

音声教材には，ゆっくり読み上げた slow バージョンと普通のスピードで読み上げた normal バージョンとの２種類がある．slow バージョンは発音の確認用だが，リスニングもシャドーイングも slow バージョンから始めた方がよいだろう．これがほぼ完璧にこなせるようになってから，normal バージョンに移行しよう．slow バージョンでは個々の単語の発音をマスターすることを目指し，normal バージョンでは単に聴き取りだけでなく，英語のリズムやある程度の長さを一息で話すことを学ぶことができる．学会発表に応用できる例文もたくさんあるので，そのリズムとともに身につけよう．**音声教材の活用の当面の目標は，完璧にシャドーイングをこなすことにする**とよいだろう．その際に文全体の意味は理解できなくてもよい．完全に聴き取って意味を理解することを次の段階では目指すとよいだろう．それぞれの例文には関連性がないうえに，文章は比較的長い．まさに単語力とリスニング力だけが頼りである．

入手できる音声ファイルは1ファイルにつき1例文になっているので，プレイリストを作って自由に並び替えたり，同じものを繰り返して聴いたりして利用できる．活用のパターンの例を示してみると以下のようになる．

1．slow 1回：発音の確認
2．normal 1回：聴き取りを短時間で行う
3．slow – slowの組み合わせ：発音の確認
4．slow – normalの組み合わせ：発音の確認と聴き取り
5．normal – normalの組み合わせ：聴き取りとその確認

例文の名前（曲名）は，文例番号とslowあるいはnormalの組み合わせになっており，曲名で並べ替えを行うとslowのファイルがnormalのファイルより前に来るようにしている．ただ415例文一度に聴くのは多すぎるかもしれないので，いくつかのグループに分けるのもいいだろう．自分の思う通りにアレンジできるのでいろいろ工夫して活用しよう．

1. slow 1回

	名前	時間
1	001_1slow	0:22
2	002_1slow	0:18
3	003_1slow	0:16
4	004_1slow	0:18
5	005_1slow	0:18
6	006_1slow	0:18

2. normal 1回

	名前	時間
1	001_2normal	0:13
2	002_2normal	0:11
3	003_2normal	0:10
4	004_2normal	0:12
5	005_2normal	0:11
6	006_2normal	0:12

4. slow-normal の組み合わせ

	名前	時間
1	001_1slow	0:22
2	001_2normal	0:13
3	002_1slow	0:18
4	002_2normal	0:11
5	003_1slow	0:16
6	003_2normal	0:10

（河本　健）

発音を身につけるためには

発音学習に発音記号は必要か？

ライフサイエンス辞書（http://lsd-project.jp/）では，英単語に発音記号を表示していない．これにはいろいろな理由があるが，その代わりとして実際の音声の収録を進めている．現在，10,000語近くの英単語の音声を聞くことができる．ライフサイエンス用語の発音の習得には，是非，この音声を活用していただきたい．

また，本書でも発音記号を示してはいないが，その代わりにすべての例文の音声教材を作製した．これらは，収録語の発音の学習にも活用できるであろう．「発音記号がないと音を聞くだけではよくわからない」と思われる方もあるかもしれないが，そのような感覚こそ本末転倒であると考えるべきではないだろうか．**発音記号は，音声を収録できない辞書での代替え手段に過ぎない．**実際，発音記号の表記方法にきちんとした国際ルールがあるわけでもなく，また，単語の発音自体にも国や地方による違いがある．発音記号は便利なものであるが，実は自転車の補助輪のようなものだ．補助輪に頼っている限り，それを離れて自走することがなかなかできない．そのため補助輪とはなるべく早く決別して，実際の音声から直接学ぶ実践的な学習方法に取り組む方がよいだろう．たとえば，日本語のフリガナとは違って英単語に発音記号を示してある本はあまり見かけない．実際に，アメリカの小学校で発音記号を教えるということはないようである．

フォニックスによる学習

では，どのように発音を学ぶのかというと，その答えの1つはフォニックス（phonix）であろう．アメリカの小学校では，低学年でフォニックスの方法を使って繰り返し発音が教えられる．英語の綴りは，日本語の仮名とは違って表音文字ではない．つまり，綴りと発音の関係が1対1ではないので，綴りを見ただけでは発音は分からないし，逆に発音を知っていても綴りは分からない．そのため正しく読むためには発音記号が必要だと我々は思いがちだが，実はそうでもないようだ．英語の綴りは表音文字ではないとはいっても，綴りに対してある程度の決まった発音のパターンがある．その綴りと発音パターンとの関係を学べば，おおよその発音は推測できる．綴りとはかけ離れた発音をする単語はごく限られた例外に過ぎない．

英語の発音のむずかしさ

英語の発音にも母音と子音とがあるが，日本語のそれらとはかなり異なっている．個々の発音については，日本でも中学校1年で必ず教えるはずであるが，多くの人は十分には習得していないので，一度，市販の発音の本で復習するのがいいだろう．日本語にない発音がたくさんあるので，意識してこれに取り組む必要がある．一方，綴りと発音の関係で戸惑うのは主に母音の発音であろう．子音に関しては，一つの綴りに対してひとつかふたつの発音しかないので，慣れれば容易に想像がつくようになる．しかし，a や o の綴りに対応する発音は3つ以上あって使い分けをマスターするのが難しい．フォニックスを使わずにこれらを学ぼうとすれば，発音記号の助けを借りることが必要になるかもしれない．しかし，発音は1文字だけで決まるのではなく，2文字以上の組み合わせが単位になることがしばしばある．3つ以上あるとはいってもそれらの限られたパターンを学んでいれば，発音記号にそれほど頼らなくても発音できる．

発音習得を効率よく進めるポイント

発音記号に頼りすぎることのもう1つの問題点は，綴りと発音との関係を学ぶことを怠るようになることにもある．ひとつひとつの単語の発音をそれぞれ個別に学習することはたいへんなことであり，発音記号に頼りすぎていると逆になかなか覚えられなくなる．

つまり，フォニックスという言葉を意識するかどうかは別として，おおよその**綴りと発音との関係とを理解したうえで，実際に聞いた音から発音を学ぶ**習慣を身につけるとかなり効率が上がるはずである．また，発音習得の際のもう１つのポイントとしては，強勢を置く母音やキーとなる子音さえきちんと発音すれば，弱く発音する母音などにはそれほど気を使わなくてもよいということが挙げられる．個々の単語の発音を忠実に再現するよりも，文全体の流れの中で発音を学ぶことの方が，正確に話すためには大切である．音声教材をたくさん真似て，よくある文章の中で発音を覚える方がずっと有用であろう．

フォニックスを学ぶ教材としては，松香洋子著『フォニックスってなんですか？』（松香フォニックス研究所）などがよいであろう．ただし，いまさらフォニックスを徹底的に学ぶ必要はないかもしれない．ルールをだいたい理解すれば，あとは音声教材を真似ることを繰り返す実践的な学習の方をむしろお勧めしたい．そして毎日，英語を聞く習慣を身につけて，学んだことを実際に活用することこそが大切である．

（河本　健）

ライフサイエンス

文例で身につける

英単語・熟語

I. 動詞の使い方

他動詞の用法には能動態と受動態とがあり，一方，自動詞には補語を必要とするものと必要としないものとがある．さまざまな動詞の用法のうち，ここでは**他動詞＋目的語**（他動詞能動態），**過去分詞＋前置詞**（他動詞受動態），**自動詞＋前置詞**の3つのパターンに絞って示す．

1 他動詞＋目的語の文例

本書の目的は英単語の語彙増強だが，単語の意味を知るだけでなく，ここに示すような**他動詞＋目的語**のよく使われる組み合わせのパターンに習熟しておくと英文を書くときに非常に役立つ．

A. 他動詞＋名詞

	文例No			用例数
assess	069	assess the effect of～	～の効果を評価する	681
	001	assess the role of～	～の役割を評価する	500
cause	002	cause an increase in～	～の増大を引き起こす	475
demonstrate	003	demonstrate the presence of～	～の存在を実証する	558
	–	demonstrate the importance of～	～の重要性を実証する	451
determine	004	determine the effect of～	～の効果を決定する	1,053
	–	determine the role of～	～の役割を決定する	814
elucidate	005	elucidate the mechanism	機構を解明する	427
evaluate	006	evaluate the effect of～	～の効果を評価する	715
	–	evaluate the role of～	～の役割を評価する	500
examine	–	examine the effect of～	～の効果を調べる	2,714
	007	examine the role of～	～の役割を調べる	1,705
	–	examine the ability of～	～の能力を調べる	418
have	008	have a role in～	～において役割を担う	580
	009	have the potential to～	～する潜在能をもつ	474
increase	–	increase the number of～	～の数を増大させる	656
	–	increase the risk of～	～のリスクを増大させる	598
	010	increase the rate of～	～の比率を増大させる	545
indicate	011	indicate the presence of～	～の存在を示す	612

	文例No			用例数
induce	012	induce the expression of ～	～の発現を誘導する	609
investigate	–	investigate the role of ～	～の役割を精査する	2,270
	013	investigate the effect of ～	～の効果を精査する	1,673
	–	investigate the mechanism	機構を精査する	912
play	201	play a role in ～	～において役割を果たす	6,552
present	382	present evidence	証拠を提示する	1,433
provide	014	provide evidence	証拠を提供する	5,103
	015	provide a mechanism	機構を提供する	739
raise	105	raise the possibility that ～	～という可能性を示唆する	1,556
reduce	–	reduce the risk of ～	～のリスクを低下させる	483
	016	reduce the number of ～	～の数を低下させる	468
regulate	017	regulate the expression of ～	～の発現を調節する	956
	–	regulate the activity of ～	～の活性を調節する	425
report	018	report the identification of ～	～の同定を報告する	585
retain	019	retain the ability to ～	～する能力を保持する	466
reveal	020	reveal the presence of ～	～の存在を明らかにする	899
study	–	study the effect of ～	～の効果を研究する	1,317
	162	study the role of ～	～の役割を研究する	872
suggest	021	suggest a role for ～	～の役割を示唆する	1,528
	–	suggest the presence of ～	～の存在を示唆する	651
	–	suggest a mechanism	機構を示唆する	631
support	–	support the hypothesis that ～	～という仮説を支持する	2,245
	022	support a model	モデルを支持する	1,073
	–	support a role for ～	～の役割を支持する	694
	236	support the idea that ～	～という考えを支持する	626
test	023	test the hypothesis that ～	～という仮説をテストする	2,855
understand	024	understand the mechanism	機構を理解する	841

001 To further investigate this possibility, we **assessed the roles of** two other proteins important for cell elongation.

- □ **further** [29,238]：副 さらに，形 さらに進んだ（far の比較級）
- □ **investigate** [30,619]：動 精査する（類 examine [39,969]，study [167,914]，test [45,529]）
- □ **possibility** [6,450]：名 可能性（類 potential [37,200]，probability [3,989]，feasibility [1,290]）
- □ **assess the role of ～** [500]：～の役割を評価する
- □ **important for ～** [7,630]：～のために重要な
- □ **elongation** [4,255]：名 伸長

訳 この可能性をさらに精査するために，我々は細胞伸長にとって重要である２つの他のタンパク質の役割を評価した

002 The bovine cell grafts in pigs **caused an increase in** antibodies detected in serum samples.

- □ **bovine** [4,868]：形 ウシの
- □ **cell** [515,640]：名 細胞
- □ **graft** [11,730]：名 移植片，移植 / 動 移植する
- □ **pig** [3,949]：名 ブタ
- □ **cause an increase in ～** [475]：～の増大を引き起こす
- □ **antibody** [36,724]：名 抗体
- □ **detect** [30,240]：動 検出する（類 find [78,775]，identify [73,456]）
- □ **detected in ～** [9,116]：～において検出される
- □ **serum sample** [784]：血清試料

訳 ブタへのウシ細胞の移植は，血清試料中に検出される抗体の増大を引き起こした

003 The results **demonstrated the presence of** a maternal autoimmune disease.

- □ **result** [158,408]：名 結果（類 consequence [7,011]，outcome [14,549]）/ 動 結果になる
- □ **demonstrate** [80,078]：動 実証する（類 document [4,359]，prove [4,555]）

demonstrate the presence of ～ [558]：～の存在を実証する

☐ **maternal** [4,162]：形 母体の，母性の

☐ **autoimmune disease** [1,624]：自己免疫疾患

訳 それらの結果は母体の自己免疫疾患の存在を実証した

004 We sought to **determine the effect of** G protein-coupled receptor kinase on bone formation in vivo.

☐ **seek**[4,216]：動 ～しようと努める（類 attempt[3,991]，try[310]，aim[4,574]），探究する（類 search [5,645]）

sought to ～ [2,674]：～しようと努めた

☐ **determine** [58,574]：動 決定する（類 define [22,878]）

determine the effect of ～ [1,053]：～の効果を決定する

☐ **G protein-coupled receptor** [2,496]：G タンパク質共役型受容体

☐ **kinase** [68,918]：名 キナーゼ，リン酸化酵素

☐ **bone formation** [574]：骨形成

☐ **in vivo** [41,680]：副 生体内で / 形 生体内の

訳 我々は，生体内での骨形成に対するGタンパク質共役型受容体キナーゼの効果を決定しようと努めた

005 These results suggest that this is a valuable model for **elucidating the mechanism** of prion conversion.

☐ **suggest** [126,371]：動 示唆する（類 imply [5,200]，mean [24,604]，indicate [738,55]）

☐ **valuable** [1,833]：形 価値ある，役立つ

☐ **model for ～** [6,728]：～のためのモデル

☐ elucidate the mechanism [427]：機構を解明する

☐ **prion** [1,907]：名 プリオン

☐ **conversion** [5,249]：名 変換

訳 これらの結果は，これがプリオン変換の機構を解明するための価値のあるモデルであることを示唆する

006

Mouse models can be useful in **evaluating the effect of** a human candidate gene mutation on an intermediate trait.

- ☐ **mouse model** [3,173]：マウスモデル
- ☐ **useful in ～ ing** [1,046]：～するのに有用である
- ☐ **evaluate the effect of ～** [715]：～の効果を評価する
- ☐ **human** [101,441]：形 ヒトの
- ☐ **candidate** [7,050]：名 候補
- ☐ **gene** [206,226]：名 遺伝子
- ☐ **mutation** [70,231]：名 変異（類 variation [10,447]）
- ☐ **intermediate** [14,897]：形 中間の
- ☐ **trait** [3,921]：名 形質

訳 マウスモデルは，中間の形質に対するヒトの候補遺伝子変異の効果を評価するのに有用でありうる

007

The aim of the current study was to **examine the role of** loud noise in acoustic neuroma etiology.

- ☐ **aim** [4,574]：名 目的（類 purpose [4,516]，objective [3,041]，goal [3,570]）／動 目的とする
- ☐ **current** [18,486]：形 現在の
- ☐ **study** [167,914]：名 研究（類 investigation [6,218]，research [8,330]）／動 研究する
- ☐ **examine the role of ～** [1,705]：～の役割を調べる
- ☐ **loud** [42]：形 騒々しい，音が大きい
- ☐ **noise** [493]：名 雑音
- ☐ **acoustic** [700]：形 聴覚の
- ☐ **neuroma** [31]：名 神経腫
- ☐ **etiology** [1,851]：名 病因，病因論（類 pathogenesis [8,673]）

訳 現在の研究の目的は，聴神経腫瘍の病因における騒々しい雑音の役割を調べることであった

008 These genes are known to **have a role in** an mRNA degradation pathway called RNA interference.

- ☐ **known to ～** [8,375]：～すると知られている
- ☐ **have a role in ～** [580]：～において役割を担う
- ☐ **mRNA** [43,887]：名 メッセンジャーRNA
- ☐ **degradation** [11,582]：名 分解（類 breakdown [1,240], decomposition [635]）
- ☐ **pathway** [59,666]：名 経路
- ☐ **called** [5,012]：～と呼ばれる（類 termed [4,067]）
- ☐ **RNA interference** [1,292]：RNA干渉

訳 これらの遺伝子は，RNA干渉といわれるメッセンジャーRNA分解経路において役割を担うということが知られている

009 These animal antiapoptotic genes **have the potential to** generate effective disease resistance in economically important crops.

- ☐ **animal** [28,747]：形 動物の，名 動物
- ☐ **antiapoptotic** [1,254]：形 抗アポトーシス性の
- ☐ **have the potential to ～** [474]：～する潜在能をもつ
- ☐ **generate** [28,107]：動 産生する（類 produce [38,705], create [6,742], synthesize [9,021]），生成する
- ☐ **effective** [14,834]：形 効果的な（類 efficacious [851], efficient [8,909]）
- ☐ **disease resistance** [412]：耐病性
- ☐ **economically** [149]：副 経済的に
- ☐ **important** [45,928]：形 重要な（類 key [13,009], vital [1,338], critical [21,712], crucial [4,757]）
- ☐ **crop** [671]：名 作物

訳 これらの動物の抗アポトーシス遺伝子は，経済的に重要な作物における効果的な耐病性を生成する潜在能を持っている

010 These mutations **increased the rate of** recombination between DNA sequences that had a high degree of sequence homology.

- □ **mutation** [70,231]：名 変異
- □ **increase the rate of ～** [545]：～の比率を増大させる
- □ **recombination** [10,831]：名 組換え
- □ **DNA sequence** [3,865]：DNA 配列
- □ **a high degree of ～** [919]：高い程度の～
- □ **sequence homology** [1,170]：配列相同性

訳 これらの変異は，高い程度の配列相同性をもつDNA配列の間の組換えの比率を増大させた

011 These observations **indicate the presence of** allergen-specific patches consisting of an unusually high proportion of surface-exposed hydrophobic residues.

- □ **observation** [14,035]：名 観察（類 detection [11,374]，discovery [3,618]，finding [35,256]）
- □ **indicate** [73,855]：動 示す（類 present [53,747]，show [148,875]，exhibit [27,339]）
- □ **indicate the presence of ～** [612]：～の存在を示す
- □ **allergen** [1,084]：名 アレルゲン
- □ **specific** [85,296]：形 特異的な
- □ **patch** [5,237]：名 パッチ / 動 パッチをあてる
- □ **consist of ～** [9,589]：～からなる（類 comprise of [1,103]，composed of [4,984]）
- □ **unusually** [935]：副 異常に，著しく（類 markedly [7,503]，extremely [2,414]）
- □ **high** [68,406]：形 高い
- □ **proportion** [5,512]：名 比率，割合（類 rate [59,576]）
- □ **surface-exposed** [364]：表面に露出した
- □ **hydrophobic** [6,484]：形 疎水性の
- □ **residue** [45,837]：名 残基

訳 これらの観察は，異常に高い割合の表面に露出した疎水性残基からなるアレルゲン特異的なパッチの存在を示す

012 In vitro, Chlamydia pneumoniae can **induce the expression of** varied molecules in infected human endothelial cells.

- □ in vitro [41,334]：副 試験管内で / 形 試験管内の（類 ex vivo [1,980] / 反 in vivo [41,680]）
- □ Chlamydia pneumoniae [165]：クラミジア肺炎菌
- □ induce the expression of ～ [609]：～の発現を誘導する
- □ varied [3,378]：多様な
- □ molecule [36,845]：名 分子
- □ infected [20,742]：感染した
- □ endothelial cell [10,488]：内皮細胞

訳 試験管内で，クラミジア肺炎菌は，感染したヒトの内皮細胞において多様な分子の発現を誘導しうる

013 We systematically **investigated the effects of** these remarkable new agents in children and adolescents.

- □ systematically [1,269]：副 系統的に
- □ investigate the effect of ～ [884]：～の効果を精査する
- □ remarkable [1,492]：形 注目すべき，顕著な（類 notable [673]，striking [2,590]，marked [6,470]，prominent [2,934]）
- □ new [34,719]：形 新しい（類 novel [30,313]）
- □ agent [18,880]：名 薬剤（類 drug [23,365]），作用物質
- □ adolescent [615]：名 青年 / 形 青年期の

訳 我々は，子供と青年におけるこれらの注目すべき新しい薬剤の効果を系統的に精査した

014 Ultrasonographic examination of the liver **provides evidence** of specific patterns of fibrosis.

- □ ultrasonographic [92]：形 超音波検査の
- □ examination [6,511]：名 検査（類 study [167,914]，research [8,330]，investigation [6,218]，test [45,529]）
- □ liver [21,312]：名 肝臓

☐ provide evidence [5,103]：証拠を提供する

☐ specific [85,296]：形 特異的な（類 special [1,078], particular [7,937]）

☐ pattern [33,282]：名 パターン，様式（類 fashion [3,575], manner [13,267]）/ 動 パターン化する

☐ fibrosis [3,793]：名 線維症

訳 肝臓の超音波検査は，線維症の特異的なパターンの証拠を提供する

015 These findings **provide a mechanism** to explain patterns of gene expression in breast cancer, colon cancer, and malignant melanoma.

☐ finding [35,256]：名 知見（類 discovery [3,618], observation [14,035]）

☐ provide a mechanism [739]：機構を提供する

☐ explain [9,556]：動 説明する（類 account for [8,259], illustrate [28,70]）

☐ pattern [33,282]：名 パターン，様式

☐ gene expression [21,171]：遺伝子発現

☐ breast cancer [8,621]：乳癌

☐ colon cancer [1,706]：結腸癌

☐ malignant [4,511]：形 悪性の

☐ melanoma [5,108]：名 メラノーマ，黒色腫

訳 これらの知見は，乳癌，結腸癌，および悪性メラノーマにおける遺伝子発現のパターンを説明する機構を提供する

016 The results indicated that vaginal epithelial application of these synthetic oligonucleotides **reduced the number of** animals that developed signs of genital herpes.

☐ indicate that ～ [58,147]：～ということを示す

☐ vaginal [1,055]：形 膣の

☐ epithelial [16,204]：形 上皮の

☐ application [10,335]：名 適用，アプリケーション

☐ synthetic [6,936]：形 合成の

☐ oligonucleotide [6,824]：名 オリゴヌクレオチド

☐ reduce [57,875]：動 低下させる（類 decrease [51,139], diminish [5,477], lower [25,400], down-regulate [2,897]） reduce the number of ～ [468]：～の数を低下させる

- □ **animal** [28,747]：名 動物 / 形 動物の
- □ **develop** [38,325]：動 発症する，開発する
- □ **sign** [2,742]：名 徴候
- □ **genital herpes** [102]：性器ヘルペス

訳 それらの結果は，これらの合成オリゴヌクレオチドの膣上皮適用は，性器ヘルペスの徴候を発症する動物の数を低下させるということを示した

017 These orphan nuclear receptors have been proposed to **regulate the expression of** detoxifying enzymes and transporters.

- □ **orphan nuclear receptor** [154]：オーファン核内受容体
- □ **propose** [20,814]：動 提唱する，提案する（類 advocate [268]，suggest [126,371]）
- □ regulate the expression of ～ [956]：～の発現を調節する
- □ **detoxify** [275]：動 解毒する
- □ **enzyme** [49,164]：名 酵素
- □ **transporter** [7,780]：名 トランスポーター

訳 これらのオーファン核内受容体は，解毒酵素とトランスポーターの発現を調節することが提唱されてきた

018 We previously **reported the identification of** a novel human protein that shuttles between the nucleus and cytoplasm.

- □ **previously** [34,828]：副 以前に（類 formerly [437]，before [16,833]）
- □ report the identification of ～ [956]：～の同定を報告する
- □ **novel** [30,313]：形 新規の，新しい（類 new [34,719]）
- □ **shuttle** [1,102]：動 シャトルする
- □ **nucleus** [19,747]：名 核
- □ **cytoplasm** [4,856]：名 細胞質

訳 我々は以前に核と細胞質の間をシャトルする新規のヒトのタンパク質の同定を報告した

019 These mutants **retained the ability to** cause lethal infections, and thus enabled us to analyze the virulence factors in a surrogate animal model.

- □ **mutant** [79,727]：名 変異体（類 variant [13,991]）/ 形 変異の
- □ **retain the ability to 〜** [466]：〜する能力を保持する
- □ **cause** [46,697]：動 引き起こす（類 result in [48,455], lead to [31,651], produce [38,705]）/ 名 原因，理由
- □ **lethal** [4,038]：形 致死的な
- □ **infection** [44,601]：名 感染
- □ **enable** [5,078]：動 可能にする
- □ **analyze** [18,249]：動 分析する，解析する（類 dissect [1,331], examine [39,969]）
- □ **virulence** [5,218]：名 病原性
- □ **factor** [103,476]：名 因子
- □ **surrogate** [988]：形 代替の / 名 代理
- □ **animal model** [3,396]：動物モデル

訳 これらの変異体は，致死的な感染を引き起こす能力を保持しており，それゆえ我々が代替の動物モデルにおける病原性因子を分析することを可能にした

020 Sequencing analysis **revealed the presence of** several known genes.

- □ **sequencing** [5,530]：名 シークエンシング，塩基配列決定
- □ **analysis** [85,671]：名 解析，分析（類 assay [38,173], examination [6,511]）
- □ **reveal** [42,563]：動 明らかにする（類 elucidate [5,037], clarify [1,417], uncover [1,208]）
  reveal the presence of 〜 [899]：〜の存在を明らかにする
- □ **several** [34,093]：形 いくつかの（類 a number of [7,385]）
- □ **known** [34,369]：既知の

訳 シークエンシング分析は，いくつかの既知の遺伝子の存在を明らかにした

021 These results **suggest a role for** these pathways in the context of normal cell proliferation.

- ☐ suggest a role for ～ [1,528]：～の役割を示唆する
- ☐ pathway [59,666]：名 経路
- ☐ in the context of ～ [2,645]：～との関連で
- ☐ normal cell [1,297]：正常細胞
- ☐ proliferation [19,120]：名 増殖

訳 これらの結果は，正常細胞増殖との関連でこれらの経路の役割を示唆する

022 These data **support a model** in which macrophage phagocytosis is coordinately regulated by both phospholipases.

- ☐ data [71,090]：名（複数形：単 datum）データ
- ☐ support a model [1,073]：モデルを支持する
- ☐ macrophage [17,484]：名 マクロファージ
- ☐ phagocytosis [1,595]：名 貪食
- ☐ coordinately [671]：副 協調的に
- ☐ regulate [57,131]：動 調節する
- ☐ phospholipase [3,237]：名 ホスホリパーゼ

訳 これらのデータは，マクロファージ貪食が両方のホスホリパーゼによって協調的に調節されるモデルを支持する

023 This study was conducted to **test the hypothesis that** inefficient metabolism in blood vessels promotes vascular disease.

- ☐ conduct [6,182]：動 行う（類 perform [21,488]，carry out [3,749]）
- ☐ test the hypothesis that ～ [2,855]：～という仮説をテストする
- ☐ inefficient [662]：形 非効率的な
- ☐ metabolism [8,783]：名 代謝
- ☐ blood vessel [837]：血管

□ **promote** [18,750]：動 促進する（類 facilitate [10,562]，accelerate [5,044]）

□ **vascular disease** [679]：血管疾患

訳 この研究は，血管における非効率的な代謝が血管疾患を促進するという仮説をテストするために行われた

024 To **understand the mechanisms** of the mosaic formation, it is of paramount importance to identify evasion mechanisms used by virulent microorganisms.

□ **understand the mechanism** [841]：機構を理解する

□ **mosaic** [1,247]：名 モザイク

□ **formation** [41,324]：名 形成（類 assembly [14,093]）

□ **of paramount importance** [46]：最も重要な

□ **identify** [73,456]：動 同定する（find [78,775]，discover [3,549]，detect [30,240]，observe [51,935]，recognize [11,391]）

□ **evasion** [403]：名 回避，逃避

□ **used by ～** [1,192]：～によって使われる

□ **virulent** [1,322]：形 病原性のある

□ **microorganism** [1,151]：名 微生物

訳 モザイク形成の機構を理解するために，病原性微生物によって使われる回避機構を同定することが最も重要である

B. 他動詞＋ that 節

文例No			用例数
025	show that ～	～ということを示す	74,382
026	indicate that ～	～ということを示す	58,147
059	suggest that ～	～ということを示唆する	96,113
027	imply that ～	～ということを示唆する	3,465
028	demonstrate that ～	～ということを実証する	45,052
029	note that ～	～ということに注目する	360
030	conclude that ～	～ということを結論する	11,673
031	estimate that ～	～ということを見積もる	510
032	reveal that ～	～ということを明らかにする	17,316
033	confirm that ～	～ということを確認する	4,189
034	we report that ～	我々は，～ということを報告する	5,672
035	we propose that ～	我々は，～ということを提唱する	6,260
036	we hypothesize that ～	我々は，～という仮説を立てる	4,074
037	we speculate that ～	我々は，～ということを推測する	710
－	we postulate that ～	我々は，～ということを仮定する	654

025 The results **showed that** increases in neuronal spike rate were accompanied by immediate decreases in tissue oxygenation.

□ **show that ～** [74,382]：～ということを示す

□ **increase** [146,670]：名 増大（類 elevation [3,909]，augmentation [739]，increment [632]，up-regulation [3,247] / 反 decrease [51,139]）.
動 増大する，増大させる

increase in ～ [35,718]：～の増大

□ **neuronal** [14,745]：形 ニューロンの，神経細胞の

□ **spike** [2,751]：名 スパイク

□ **rate** [59,576]：名 比率，速度，割合

□ **accompany** [6,903]：動 伴う（類 follow [44,620]，associate [82,102]） accompanied by ～ [4,941]：～に伴われる，～を伴う

☐ immediate [3,094]：形 即時の

☐ decrease [51,139]：名 低下，減少（類 reduction [23,591]，fall [2,820]，attenuation [2,259]），動 低下する，低下させる

decrease in ～ [13,282]：～の低下

☐ tissue [45,858]：名 組織

☐ oxygenation [905]：名 酸素添加，酸素負荷

訳 それらの結果は，ニューロンのスパイク比率の増大は組織の酸素添加の即時の低下を伴うということを示した

026 The result **indicates that** PS1 mutations modulate intracellular calcium signaling pathways.

☐ indicate that ～ [58,147]：～ということを示す

☐ mutation [70,231]：名 変異

☐ modulate [13,347]：動 調節する（類 regulate [57,131]，control [78,085]，mediate [71,309]）

☐ intracellular [20,566]：形 細胞内の（反 extracellular [16,514]）

☐ calcium [17,580]：名 カルシウム

☐ signaling pathway [9,274]：シグナル経路，情報伝達経路

訳 その結果は，PS1 変異が細胞内カルシウムシグナル経路を調節するということを示す

027 The findings **imply that** chemokine networks serve important functions at the maternal-fetal interface.

☐ finding [35,256]：名 知見

☐ imply [5,200]：動 示唆する，意味する（類 mean [24,604]，indicate [73,855]，suggest [126,371]）

imply that ～ [3,465]：～ということを示唆する

☐ chemokine [6,964]：名 ケモカイン

☐ network [9,624]：名 ネットワーク

☐ serve [9,796]：動 働く，役立つ（類 function [95,000]，act [19,853]，behave [1,367]，work [12,304]），務める

□ **important** [45,928]：形 重要な

□ **function** [92,343]：名 機能 / 動 機能する

□ **maternal** [4,162]：形 母体の

□ **fetal** [9,274]：形 胎児性の

□ **interface** [9,274]：名 界面

訳 その知見は，ケモカインのネットワークが母体と胎児の境界において重要な機能を務めるということを意味している

028 Data presented here **demonstrate that** biofilm formation was evident in all specimens.

□ **present** [53,747]：動 提示する，示す（類 show [148,875]，indicate [73,855]，exhibit [27,339]）/ 形 存在する，現在の

□ **demonstrate that ～** [45,052]：～ということを実証する

□ **biofilm** [1,631]：名 バイオフィルム，生物膜

□ **formation** [41,324]：名 形成

□ **evident** [2,899]：形 明らかな（類 apparent [8,392]，clear [5,786]，obvious [1,088]）

□ **specimen** [5,996]：名 検体（類 sample [20,440]，preparation [5,341]）

訳 ここに示されたデータは，バイオフィルム形成がすべての検体において明らかであったことを実証する

029 Physicians should **note that** many months usually pass between the diagnosis of cancer and the occurrence of complications.

□ **physician** [4,789]：名 内科医，医師（類 doctor [207]）

□ **note** [3,677]：動 注目する，述べる

note that ～ [360]：～ということに注目する

□ **usually** [3,166]：副 普通，通常（類 generally [5,546]，commonly [4,265]）

□ **pass** [1,341]：動 経過する，通過する

□ **diagnosis** [8,763]：名 診断（類 assessment [5,256]）

□ **cancer** [45,184]：名 癌（類 carcinoma [45,184]，tumor [61,098]）

□ **occurrence** [2,697]：名 発生（類 appearance [3,367]，emergence [1,340]）

□ **complication** [5,903]：名 合併症

訳 内科医は，癌の診断と合併症の発生の間に普通は数ヶ月が経過するということに注目すべきである

030 We **conclude that** patients with some endogenous insulin secretory capacity do not depend on insulin for immediate survival.

□ **conclude** [12,191]：動 結論する

conclude that ～ [11,673]：～ということを結論する

□ **patient with ～** [43,273]：～の患者

□ **endogenous** [13,368]：形 内在性の（類 intrinsic [5,126] / 反 exogenous [5,541]）

□ **insulin** [20,353]：名 インスリン

□ **secretory** [3,789]：形 分泌性の

□ **capacity** [8,285]：名 能力（類 ability [29,720]，capability [1,731]，competence [1,025]，potential [37,200]），容量

□ **depend** [12,099]：動 依存する

depend on ～ [9,937]：～に依存する

□ **immediate** [3,094]：形 即時の，当面の

□ **survival** [28,139]：名 生存，生存率（類 viability [3,419]）

訳 我々は，いくらかの内在性インスリン分泌能を持つ患者は，当面の生存のためにはインスリンに依存しないと結論づける

031 Authorities **estimate that** at least 3 million children suffer from lead poisoning.

□ **authority** [117]：名 当局，権威（者）

□ **estimate** [13,499]：動 見積もる（類 expect [6,909]，presume [1,195]，assume [2,937]）

estimate that ～ [510]：～ということを見積もる

□ **at least** [16,456]：少なくとも

□ **million** [2,566]：名 100 万

□ **suffer** [1,127]：動 患う，受ける

suffer from ～ [517]：～を患う

☐ **lead** [9,294]：名 鉛 / 動 つながる

☐ **poisoning** [311]：名 中毒

訳 当局は，少なくとも300万人の子供達が鉛中毒を患っていると見積もっている

032 Pathologic studies **reveal that** the most likely primary site of origin includes the colon and kidneys.

☐ **pathologic** [1,842]：形 病理学的な

☐ **reveal that ～** [17,316]：～ということを明らかにする

☐ **most likely** [2,278]：最もありそうな

☐ **primary** [28,650]：形 原発の，一次の，主要な

☐ **site** [99,321]：名 部位（類 area [5,037]，location [8,396]，locus [18,724]，region [77,257]）

☐ **origin** [7,942]：名 起源，由来（類 source [10,013]）

☐ **include** [59,168]：動 含む（類 contain [71,403]，involve [49,378]）

☐ **colon** [4,699]：名 大腸，結腸

☐ **kidney** [10,654]：名 腎臓

訳 病理学的研究は，起源の最もありそうな原発部位が大腸と腎臓を含むということを明らかにする

033 No epidemiologic studies have yet **confirmed that** women with migraine headaches have a greater risk of stroke.

☐ **epidemiologic** [1,220]：形 疫学の

epidemiologic study [520]：疫学調査，疫学研究

☐ **confirm that ～** [4,189]：～ということを確認する

☐ **migraine headache** [26]：片頭痛

☐ **greater** [19,331]：形（great の比較級）より大きな

☐ **risk** [37,608]：名 リスク，危険

☐ **stroke** [4,568]：名 脳卒中，発作

訳 偏頭痛を持つ女性は，脳卒中のより大きなリスクを持つということを確認した疫学研究はまだない

034 **We report that** the majority of cells in the hypothalamic circadian pacemaker are lost in patients with senile dementia.

- ☐ **we report that ～** [5,672]：我々は，～ということを報告する
- ☐ **the majority of ～** [4,521]：～の大部分
- ☐ **hypothalamic** [1,849]：形 視床下部の
- ☐ **circadian** [2,709]：形 概日性の
- ☐ **pacemaker** [886]：名 ペースメーカー
- ☐ **lose** [4,468]：動（過 / 過分 lost）失う（類 abolish [7,419]，delete [3,910]）
- ☐ **patient with ～** [43,273]：～の患者
- ☐ **senile dementia** [6]：老年認知症

訳 我々は，視床下部の概日性ペースメーカーの細胞の大部分が老年認知症の患者において失われているということを報告する

035 **We propose that** these genes are members of MAP kinase signaling pathways.

- ☐ **we propose that ～** [6,260]：我々は，～ということを提唱する
- ☐ **member** [21,048]：名 メンバー，構成因子
- ☐ **MAP kinase** [3,943]：MAP キナーゼ
- ☐ **signaling pathway** [9,274]：シグナル経路

訳 我々は，これらの遺伝子がMAPキナーゼシグナル経路のメンバーであると提唱する

036 **We hypothesized that** the mechanism of genomic integration may be similar to transposition.

- ☐ **hypothesize** [6,370]：動 仮説を立てる（類 postulate [2,585]，speculate [1,149]，assume [2,937]）
  we hypothesize that ～ [4,074]：我々は，～という仮説を立てる
- ☐ **mechanism** [70,869]：名 機構（類 machinery [3,238]，mode [6,316]）
- ☐ **genomic** [11,843]：形 ゲノムの
- ☐ **integration** [4,162]：名 組込み

☐ **similar to ～** [17,306]：～に類似している

☐ **transposition** [1,168]：名 遺伝子転位，転位

訳 我々は，ゲノム組込みの機構は遺伝子転位に類似しているかもしれないという仮説を立てた

037 **We speculate that** such dysfunction may be relevant to the mutation.

☐ **speculate** [1,149]：動 推測する（類 postulate [2,585]，presume [1,195]，assume [2,937]，estimate [13,499]）

we speculate that ～ [710]：我々は，～ということ推測する

☐ **dysfunction** [6,236]：名 機能障害（類 malfunction [101]），機能不全，障害（類 impairment [3,806]，disorder [16,726]，disturbance [970]，damage [13,052]）

☐ **relevant** [5,577]：形 関連する（類 relative [21,076]，related [36,580]）

relevant to ～ [1,258]：～に関連する

☐ **mutation** [70,231]：名 変異

訳 我々は，そのような機能障害はその変異に関連するかもしれないと推測する

C. 他動詞＋ whether 節

文例No			用例数
083	determine whether ～	～かどうかを決定する	7,898
038	investigate whether ～	～かどうかを精査する	2,515
039	examine whether ～	～かどうかを調べる	1,800
–	test whether ～	～かどうかをテストする	1,856
040	ask whether ～	～かどうかを問う	581
041	assess whether ～	～かどうかを評価する	621
042	evaluate whether ～	～かどうかを評価する	464
043	address whether ～	～かどうかに取り組む	209

038 The aim of this study was to **investigate whether** T cells infected with HIV are more susceptible to Fas-induced death.

□ **aim** [4,574]：名 目的 / 動 目的とする

□ **investigate whether ～** [2,515]：～かどうかを精査する

□ **T cell** [56,789]：T 細胞

□ **infect** [22,659]：動 感染させる（infect A with B：A を B に感染させる）

□ **infected with ～** [4,538]：～に感染した

□ **HIV** [29,027]：名 ヒト免疫不全ウイルス（human immunodeficiency virus）

□ **susceptible** [4,520]：形 感受性の

susceptible to ～ [2,430]：～に感受性の

□**～ -induced** [20,000]：～に誘導される

□ **death** [26,658]：名 死

訳 この研究の目的は，HIV に感染した T 細胞が Fas に誘導される死に対しより感受性であるかどうかを精査することであった

039 This study **examined whether** the adverse effects of prenatal exposure to tobacco on lung development are limited to the last weeks of gestation.

- □ **examine whether 〜** [1,800]：〜かどうかを調べる
- □ **adverse effect** [1,014]：有害作用，副作用
- □ **prenatal** [841]：形 出生前の
- □ **exposure to 〜** [6,891]：〜への曝露
- □ **tobacco** [2,480]：名 タバコ
- □ **lung** [19,948]：名 肺
- □ **development** [48,509]：名 発生，発症，開発
- □ **limit** [21,272]：動 制限する（類 restrict [9,651]）/ 名 限界
- □ **last** [3,443]：形 最後の，動 続く
- □ **gestation** [1,216]：名 妊娠（類 pregnancy [3,407]），妊娠期間

訳 この研究は，タバコへの出生前曝露の肺発生に対する有害作用が妊娠の最後の数週間に限られるかどうかを調べた

040 In this study, we **asked whether** distinct immunochemical reactions might occur after xenotransplantation of the lung.

- □ **ask** [1,328]：動 問う
  ask whether 〜 [581]：〜かどうかを問う
- □ **distinct** [22,633]：形 別個の，明らかに異なる（類 different [48,428]，dissimilar [331]，disparate [647]，discrete [3,058]）
- □ **immunochemical** [157]：形 免疫化学的な
- □ **reaction** [31,436]：名 反応（類 response [104,667]）
- □ **occur** [42,905]：動 起こる，生じる（類 take place [1,287]，arise [6,458]，emerge [4,376]）
- □ **xenotransplantation** [284]：名 異種移植

訳 この研究において，我々は別個の免疫化学的反応が肺の異種移植の後に起こるかどうかを問うた

041 The studies reported here **assess whether** a defective oxidative defense may contribute to Down's syndrome.

- □ **report** [54,972]：動 報告する（類 describe [25,975], document [4,359]）/ 名 報告
- □ **here** [51,283]：副 ここで（類 herein [1,012]）
- □ **assess whether ～** [621]：～かどうかを評価する
- □ **defective** [7,616]：形 欠陥のある，欠損した
- □ **oxidative** [6,890]：形 酸化的な
- □ **defense** [3,423]：名 防御（類 protection [8,224]）
- □ **contribute** [23,342]：動 寄与する，一因となる
  contribute to ～ [19,994]：～に寄与する（類 result in [48,455], lead to [31,651], cause [46,697], produce [38,705]）
- □ **Down's syndrome** [170]：ダウン症候群
- □ **syndrome** [12,784]：名 症候群

訳 ここで報告される研究は，欠陥のある酸化防御がダウン症候群に寄与するかどうかを評価する

042 Further studies are required to **evaluate whether** cellular uptake of DNA is a significant barrier to efficient transfection in vivo.

- □ **further** [29,238]：形 さらに進んだ（far の比較級）/ 副 さらに
- □ **required to ～** [4,754]：～するために必要とされる
- □ **evaluate whether ～** [464]：～かどうかを評価する
- □ **cellular** [29,968]：形 細胞の，細胞性の
- □ **uptake** [11,299]：名 取り込み（類 incorporation [4,734]）
- □ **significant** [43,571]：形 重要な（類 important [45,928], marked [6,470], prominent [2,934]），有意な，著しい
- □ **barrier** [5,355]：名 障害，障壁（類 obstacle [554], disturbance [970], difficulty [1,608]）
- □ **efficient** [8,909]：形 効率的な（類 effective [14,834], efficacious [851]）
- □ **transfection** [5,761]：名 形質移入，トランスフェクション（類 transformation [7,398]）

☐ **in vivo** [41,680]：副 生体内で / 形 生体内の（反 in vitro [41,334]，ex vivo [1,980]）

訳 さらに進んだ研究が，DNAの細胞取り込みが生体内で効率的な形質移入に対する重要な障壁であるかどうかを評価するために必要とされる

043 Few animal studies have been carried out to **address whether** adolescent nicotine exposure exerts unique or lasting effects on brain structure or function.

☐ **few** [7,261]：形 ほとんどない

☐ **animal study** [344]：動物試験，動物実験

☐ **carry out** [3,749]：行う，実行する（類 perform [10,318]，conduct [2,976]）

☐ **address** [6,396]：動 取り組む，検討する

address whether ～ [209]：～かどうかに取り組む

☐ **adolescent** [615]：形 青年期の / 名 青年

☐ **nicotine** [1,742]：名 ニコチン

☐ **exposure** [16,765]：名 曝露

☐ **exert** [3,821]：動 及ぼす，発揮する

☐ **unique** [12,913]：形 独特の，ユニークな

☐ **lasting** [1,278]：形 永続する，続く

☐ **effect on ～** [23,839]：～に対する影響

☐ **brain** [28,011]：名 脳

☐ **structure** [64,485]：名 構造

☐ **function** [92,343]：名 機能 / 動 機能する

訳 青年期のニコチン曝露が脳の構造や機能に対する独特のあるいは永続する影響を及ぼすかどうかに取り組むために行われた動物試験はほとんどない

D. 他動詞＋ to *do*

to 不定詞は他動詞の目的語になる場合と自動詞の補語（appear to *do* など）になる場合とがあるので注意が必要である．

文例No			用例数
004	sought to～（seek to～）	～しようと努めた（～しようと努める）	2,674
044	we attempt to～	我々は，～しようと試みる	256
－	try to～	～しようとする	215

044 We **attempted to** identify risk factors for breast cancer.

- □ **attempt**［3,991］：動 試みる（類 try［310］，seek［4,216］）/ 名 試み
  we attempt to ～［256］：我々は，～しようと試みる
- □ **identify**［73,456］：動 同定する
- □ **risk factor**［6,751］：危険因子，リスク因子
- □ **breast cancer**［8,621］：乳癌

訳 我々は，乳癌のリスク因子を同定しようと試みた

2 他動詞（過去分詞）＋前置詞／ that 節の文例

論文では受動態の用例が非常に多い．むやみに受動態を使うことは避けるべきであるが，受動態が使われるパターンは習得しておきたいものである．受動態の過去分詞のあとには前置詞が続くことが多く，これが使い方の大きなポイントとなる．どの前置詞が続くかは単語ごとに決まった組み合わせがあるので，ここではよく使われる**過去分詞＋前置詞**に焦点を当てる．

A. 他動詞（過去分詞）＋前置詞＋名詞

文例No			用例数
045	learned about ～	～について知られている	104
–	considered as ～	～であると考えられる	438
046	regarded as ～	～であるとみなされる	367
047	viewed as ～	～であるとみなされる	367
048	aimed at ～ ing	～することを目的とする	717
174	induced by ～	～によって誘導される	13,321
–	mediated by ～	～によって仲介される	11,193
049	followed by ～	あとに～を伴う	10,013
346	caused by ～	～によって引き起こされる	8,510
050	required for ～	～に必要とされる	29,167
100	used for ～	～のために使われる	3,818
051	observed for ～	～に対して観察される	3,221
052	derived from ～	～に由来する	11,249
265	obtained from ～	～から得られる	4,871
053	inferred from ～	～から推論される	374
048	involved in ～	～に関与する	25,574
002	detected in ～	～において検出される	9,116
054	implicated in ～	～に関与する	7,864
055	located in ～	～に位置する	3,715
235	based on ～	～に基づいた	19,105

文例No			用例数
056	expressed on ～	～において発現される	1,962
057	performed on ～	～に対して行われる	1,719
－	located on ～	～に位置する	1,356
086	compared to ～	～と比較される，～と比較して	10,768
261	related to ～	～に関連する	10,949
228	associated with ～	～と関連する	54,439
058	compared with ～	～と比較される	32,460
059	correlated with ～	～と相関する	9,967
058	treated with ～	～によって治療される，によって処理される	10,055
060	infected with ～	～を感染させられる，～に感染した	4,538

045 Much has been **learned about** the underlying mechanisms for adverse effects of chemotherapy.

- □ **much** [10,556]：名 多くのこと / 形 多くの / 副 非常に
- □ **learn** [1,427]：動 知る，学習する（類 know [34,957]，understand [12,582]）
  - **learned about ～** [104]：～について知られている
- □ **underlying mechanism** [808]：根底にある機構
- □ **adverse effect** [1,014]：有害作用
- □ **chemotherapy** [5,426]：化学療法

訳 化学療法の有害作用の根底にある機構について多くが知られている

046 The suprachiasmatic nucleus is **regarded as** the main mammalian circadian pacemaker.

- □ **suprachiasmatic nucleus** [247]：視交叉上核
- □ **regard** [3,713]：動 みなす（類 consider [8,440]，view [4,519]，think [8,004]）
  - **regarded as ～** [367]：～であるとみなされる
- □ **main** [3,667]：形 主要な（類 major [30,348]，dominant [11,905]，predominant [2,685]）
- □ **mammalian** [15,396]：形 哺乳類の

☐ **circadian** [2,709]：形 概日性の

☐ **pacemaker** [886]：名 ペースメーカー

訳 視交叉上核は，哺乳類の主要な概日性ペースメーカーとみなされている

047 The prediction of protein structure is **viewed as** a great challenge for scientists.

☐ **prediction** [5,264]：名 予測，予想（類 expectation [900]，anticipation [154]）

☐ **protein structure** [1,203]：タンパク質構造

☐ **view** [4,519]：動 みなす（類 regard [3,713]，consider [8,440]），名 見解

viewed as ～ [367]：～であるとみなされる

☐ **great** [2,183]：形 大きな（類 large [27,716]，considerable [2,681]，significant [43,571]）

☐ **challenge** [10,236]：名 挑戦 / 動 挑戦する，曝露する

☐ **scientist** [440]：名 科学者

訳 タンパク質構造の予測は，科学者にとって大きな挑戦であるとみなされている

048 This study is **aimed at** identifying the signaling pathways involved in these events.

☐ **aimed at ～ ing** [717]：～することを目的とする

☐ **identify** [73,456]：動 同定する

☐ **signaling pathway** [9,274]：シグナル経路

☐ **involve** [49,378]：動 関与させる

involved in ～ [25,574]：～に関与する

☐ **event** [24,512]：名 事象（類 case [25,633]，incident [1,189]），現象（類 phenomenon [3,643]）

訳 この研究は，これらの事象に関与するシグナル経路を同定することを目的とする

049 Industrial aerosol exposure leads to headache, drowsiness, nausea, and vomiting, **followed by** a latent period of 1 to 5 days before the onset of additional symptoms.

- □ **industrial** [329]：形 工業の，産業の
- □ **aerosol** [728]：名 エアロゾル
- □ **exposure** [16,765]：名 曝露
- □ **lead** [37,185]：動 つながる / 名 鉛
  lead to ～ [31,651]：～につながる
- □ **headache** [456]：名 頭痛
- □ **drowsiness** [13]：名 眠気
- □ **nausea** [424]：名 悪心
- □ **vomiting** [354]：名 嘔吐
- □ **followed by ～** [10,013]：あとに～を伴う
- □ **latent** [2,303]：形 潜伏した
- □ **period** [14,240]：名 期間（類 term [18,638]）
- □ **onset** [9,566]：名 開始，発症（類 initiation [10,562]，start [6,641]）
- □ **symptom** [9,302]：名 症状（類 manifestation [1,760]，sign [2,848]，indication [1,386]）

訳 工業エアロゾル曝露は頭痛，眠気，悪心および嘔吐につながり，さらなる症状の発生の前に 1 から 5 日間の潜伏期を伴う

050 Molecular interactions are **required for** normal signal transduction across the lipid bilayer.

- □ **molecular** [36,050]：形 分子の
  molecular interaction [429]：分子間相互作用
- □ **required for ～** [29,167]：～に必要とされる
- □ **normal** [44,556]：形 正常な
- □ **signal transduction** [7,006]：シグナル伝達
- □ **lipid bilayer** [1,093]：脂質二重層

訳 分子の相互作用が，脂質二重層を越える正常なシグナル伝達に必要とされる

051 The patient should be **observed for** evidence of proximal colonic bleeding.

- □ **patient** [145,663]：名 患者
- □ **observed for ～** [3,221]：～に対して観察される
- □ **evidence** [35,499]：名 証拠（類 proof [566], hallmark [1,363], demonstration [2,008]），動 立証する
- □ **proximal** [6,711]：形 近位の
- □ **colonic** [1,426]：形 大腸の
- □ **bleeding** [1,847]：名 出血（類 hemorrhage [1,415]）

訳 患者は，近位の大腸の出血の証拠を求めて観察されるべきである

052 This virus-like particle was discovered in an infectious stool filtrate **derived from** an outbreak of gastroenteritis in a preschool.

- □ **virus** [47,464]：名 ウイルス
- □ **particle** [8,576]：名 粒子
- □ **discover** [3,549]：動 発見する（類 find [78,775], identify [73,456], detect [30,240]）
- □ **infectious** [4,116]：形 感染性の
- □ **stool** [640]：名 糞便（類 feces [218]）
- □ **filtrate** [147]：名 濾液
- □ **derive** [28,156]：動 引き出す，由来する
  - **derived from ～** [11,249]：～に由来する（類 originate from [1,236], arise from [3,170], stem from [302], come from [703]）
- □ **outbreak** [1,458]：名 大流行，集団発生
- □ **gastroenteritis** [244]：名 胃腸炎
- □ **preschool** [78]：名 幼稚園

訳 このウイルス様粒子は，幼稚園における胃腸炎の集団発生に由来する感染性糞便濾液中に発見された

053 Several potential binding sites can be **inferred from** functional mapping.

☐ **potential** [37,200]：形 可能な，潜在的な（類 possible [14,443]，feasible [1,123]），名 潜在能

☐ **binding site** [18,396]：結合部位

☐ **infer** [1,727]：動 推論する（類 expect [6,909]，predict [20,165]，presume [1,195]，speculate [1,149]，postulate [2,585]，assume [2,937]，deduce [1,828]）

inferred from 〜 [374]：〜から推論される

☐ **functional** [34,575]：形 機能的な

☐ **mapping** [6,037]：形 地図作成，マッピング

訳 いくつかの可能な結合部位が，機能マッピングから推論されうる

054 Several growth factors **implicated in** the stimulation of fibroblasts are capable of enhancing the production and secretion of extracellular matrix proteins.

☐ **growth factor** [18,400]：増殖因子，成長因子

☐ **implicated in 〜** [7,864]：〜に関与する（類 involved in [25,574]，engaged in [297]，participate in [5,828]）

☐ **stimulation** [21,013]：名 刺激（類 irritation [100]）

☐ **fibroblast** [12,722]：名 線維芽細胞

☐ **capable of 〜 ing** [7,651]：〜できる，〜する能力がある（類 able to 〜 [9,352] / 反 incapable of 〜 [611]）

☐ **enhance** [32,256]：動 増強する（類 potentiate [2,958]，augment [3,582]，reinforce [776]，facilitate [10,562]）

☐ **production** [27,611]：名 産生（類 generation [10,962]）

☐ **secretion** [12,758]：名 分泌（類 discharge [2,812]，release [23,508]），分泌物

☐ **extracellular matrix** [3,238]：細胞外基質

☐ **protein** [302,823]：名 タンパク質

訳 線維芽細胞の刺激に関与するいくつかの増殖因子は，細胞外基質タンパク質の産生と分泌を増大させることができる

055 The lesions of lymphoma are often found to be **located in** the white matter adjacent to the ventricles.

- □ **lesion** [18,776]：名 病変（部）/ 動 破壊する
- □ **lymphoma** [5,229]：名 リンパ腫
- □ **often** [9,004]：副 しばしば（類 frequently [5,561]）
- □ **found to** ～ [15,859]：～することが見つけられる
- □ **locate** [12,854]：動 位置づける
  - **located in** ～ [3,715]：～に位置する（類 lie in [669]）
- □ **white matter** [1,210]：白質
- □ **adjacent** [6,670]：形 隣接する（類 close [5,537], proximate [154]）
  - **adjacent to** ～ [2,547]：～に隣接する
- □ **ventricle** [1,586]：名 脳室，心室

訳 リンパ腫の病変部は，しばしば脳室に隣接する白質に位置することが見つけられる

056 This protein product **expressed on** the surface of tumor cells structurally appears to be a tyrosine kinase receptor analogous to the epidermal growth factor receptor.

- □ **product** [24,771]：名 産物
- □ **expressed on** ～ [1,962]：～において発現される
- □ **surface** [37,833]：名 表面
- □ **tumor cell** [7,332]：腫瘍細胞
- □ **structurally** [3,502]：副 構造的に
- □ **appear** [28,873]：動 思われる
  - **appear to** ～ [21,779]：～するように思われる
- □ **tyrosine kinase receptor** [277]：チロシンキナーゼ受容体
- □ **analogous** [2,353]：形 類似の（類 similar [45,428], comparable [6,413], equivalent [4,744], homologous [8,890]）
  - **analogous to** ～ [1,160]：～に類似する
- □ **epidermal growth factor receptor** [1,124]：上皮増殖因子受容体

訳 腫瘍細胞の表面で発現されるこのタンパク質産物は，構造的に上皮増殖因子受容体に類似するチロシンキナーゼ受容体であると思われる

057 Specialized tests should be **performed on** an appropriately prepared tumor biopsy to diagnose responsive endometrial cancer.

- □ **specialize** [2,236]：動 特殊化する
- □ **test** [45,529]：名 テスト，検査（類 examination [6,511]，assessment [5,256]），動 テストする，検査する
- □ **performed on ～** [1,719]：～に対して行われる
- □ **appropriately** [784]：副 適切に（類 adequately [549]）
- □ **prepare** [6,376]：動 調製する，作成する（類 construct [15,084]）
- □ **tumor** [61,098]：名 腫瘍（類 cancer [45,184]）
- □ **biopsy** [5,466]：名 生検，バイオプシー
- □ **diagnose** [4,479]：動 診断する
- □ **responsive** [6,082]：形 応答性の（類 reactive [8,104]）
- □ **endometrial** [782]：形 子宮内膜の

訳 特殊化されたテストが，応答性の子宮内膜癌を診断するために適切に調製された腫瘍生検に対して行われるべきである

058 Five-year survival of interferon-treated patients was **compared with** that of those treated with chemotherapy.

- □ **survival** [28,139]：名 生存，生存率
- □ **interferon** [6,201]：名 インターフェロン
- □ **～ -treated** [10,000]：～に処理された
- □ **compare** [58,415]：動 比較する
  - **compared with ～** [32,460]：～と比較される
- □ **treated with ～** [10,055]：～で治療される，～によって処理される
- □ **chemotherapy** [5,426]：名 化学療法

訳 インターフェロンで治療された患者の5年生存率が，化学療法によって治療された患者に対するそれと比較された

059 These results suggest that fat intake is **correlated with** the risk for colorectal cancer.

- □ **suggest that ～** [96,112]：～ということを示唆する

□ **fat** [6,421]：名 脂肪

□ **intake** [5,902]：名 摂取

□ **correlate** [21,378]：動 相関させる

correlated with ～ [9,967]：～と相関する（類 related to [10,949]，associated with [54,439]，inked to [5,539]）

□ **risk for ～** [4,282]：～のリスク

□ **colorectal cancer** [1,720]：結腸直腸癌，大腸癌

訳 これらの結果は，脂肪摂取が結腸直腸癌のリスクと相関するということを示唆する

060 Animal-associated opportunistic infections were reported to be found among persons **infected with** the human immunodeficiency virus.

□ **～-associated** [10,000]：～と関連する

□ **opportunistic infection** [269]：日和見感染

□ **reported to ～** [2,484]：～すると報告される

□ **find** [78,775]：動（過 / 過分 found）見つける（類 discover [3,549]，identify [73,456]，detect [30,240]）

□ **person** [5,218]：名 人（類 individual [24,178]）

□ **infected with ～** [4,538]：～を感染させられる，～に感染した

□ **human immunodeficiency virus** [5,064]：ヒト免疫不全ウイルス

訳 動物と関連する日和見感染は，ヒト免疫不全ウイルスに感染した人々の中で見つかると報告された

B. 他動詞（過去分詞）＋ to *do*

文例No			用例数
061	expected to ～	～すると予想される	1,448
062	predicted to ～	～すると予測される	2,555
063	presumed to ～	～すると推定される	431
064	postulated to ～	～すると仮定される	748
－	proposed to ～	～すると提唱される	2,688
－	hypothesized to ～	～すると仮定される	668
065	thought to ～	～すると考えられている	6,707
066	believed to ～	～すると信じられている	2,472
067	considered to be ～	～であると考えられる	770
068	known to ～	～すると知られている	8,375
285	shown to ～	～することが示される	17,485
121	found to ～	～することが見つけられる	15,859
069	used to ～	～するために使われる	26,204
070	designed to ～	～するために計画される	3,933
071	undertaken to ～	～するために着手される	1,050

061 The numbers are **expected to** increase by 20% in the next decade.

☐ **expect** [6,909]：動 予想する（類 predict [20,165], presume [1,195], infer [1,727], assume [2,937], postulate [2,585], speculate [1,149]），期待する

expected to ～ [1,448]：～すると予想される

☐ **increase** [146,670]：動 増大する，増大させる，名 増大

☐ **next** [2,355]：形 次の（類 following [20,514]），副 次に

☐ **decade** [2,311]：名 十年

訳 それらの数は，次の10年間に20％増大すると予想される

062

The gene is **predicted to** encode an integral membrane protein.

☐ **predict** [20,165]：動 予測する，予想する（類 expect [6,909]，presume [1,195]，infer [1,727]，assume [2,937]，speculate [1,149]，postulate [2,585]，deduce [1,828]）

predicted to ～ [2,555]：～すると予測される

☐ **encode** [37,774]：動 コードする（類 code [9,124] / 反 decode [365]）

☐ **integral membrane protein** [819]：膜内在性タンパク質

訳 その遺伝子は，膜内在性タンパク質をコードすると予測される

063

These responses are **presumed to** be modulated by protein-protein interactions.

☐ **response** [104,667]：名 反応（類 reaction [31,436]），応答

☐ **presume** [1,195]：動 推定する（類 predict [20,165]，expect [6,909]，infer [1,727]，assume [2,937]，speculate [1,149]，postulate [2,585]，estimate [13,499]）

presumed to ～ [431]：～すると推定される

☐ **modulate** [13,347]：動 調節する

☐ **protein-protein interaction** [2,190]：タンパク質ータンパク質相互作用

訳 これらの反応は，タンパク質ータンパク質相互作用によって調節されると推定される

064

Aβ accumulation has been **postulated to** contribute to the pathogenesis of Alzheimer's disease.

☐ **accumulation** [13,180]：名 蓄積（類 deposition [3,189]，deposit [1,172]）

☐ **postulate** [2,585]：動（自明のこととして）仮定する（類 hypothesize [6,370]，speculate [1,149]，expect [6,909]，predict [20,165]，presume [1,195]，infer [1,727]），みなす

postulated to ～ [748]：～すると仮定される

☐ **contribute to ～** [19,994]：～に寄与する

☐ **pathogenesis** [8,673]：名 病因（類 etiology [1,851]）

☐ **Alzheimer's disease** [2,946]：アルツハイマー病

訳 Aβ沈着は，アルツハイマー病の病因に寄与すると仮定されている

065

Cyclin D1 is **thought to** be a key regulator involved in cell cycle progression.

☐ **cyclin** [12,033]：名 サイクリン

☐ **think** [8,004]：動（過／過分 thought）考える（類 consider [8,440]，regard [3,713]，believe [3,698]）

thought to ～ [6,707]：～すると考えられている

☐ **key** [13,044]：形 鍵となる，重要な（類 important [45,928]，critical [21,712]，crucial [4,757]，vital [1,338]），名 鍵

☐ **regulator** [11,450]：名 制御因子

☐ **involved in ～** [25,574]：～に関与する

☐ **cell cycle progression** [2,039]：細胞周期進行

訳 サイクリンD1は，細胞周期進行に関与する決定的に重要な調節因子であると考えられている

066

The vitamin D receptor is **believed to** mediate this activity.

☐ **vitamin** [5,290]：名 ビタミン

☐ **receptor** [125,789]：名 受容体

☐ **believe** [3,698]：動 信じる（類 think [8,004]，accept [1,866]，appreciate [452]，regard [3,713]）

believed to ～ [2,472]：～すると信じられている

☐ **mediate** [71,309]：動 仲介する（類 modulate [13,347]，control [78,085]，regulate [57,131]）

☐ **activity** [33,998]：名 活性，活動

訳 ビタミンD受容体は，この活性を仲介すると信じられている

067

Acetylation of histones is **considered to be** a critical step in transcriptional regulation.

☐ **acetylation** [2,333]：名 アセチル化

☐ **histone** [8,536]：名 ヒストン

☐ considered to be ～ [770]：～であると考えられる

☐ critical [21,712]：動 決定的に重要な（類 crucial [4,757]，definitive [1,252]，important [45,928]，conclusive [176]）

☐ step in ～ [4,731]：～における段階

☐ transcriptional regulation [1,633]：転写調節

訳 ヒストンのアセチル化は，転写調節における決定的に重要な段階であると考えられる

068

Wnt signaling is **known to** be involved in early steps of neural crest development.

☐ signaling [42,594]：名 シグナル伝達

☐ known to ～ [8,375]：～すると知られている

☐ involved in ～ [25,574]：～に関与する（類 implicated in [7,864]，engaged in [297]，participate in [5,828]）

☐ early [31,638]：形 早期の，初期の（類 initial [12,861]，original [2,348]），副 早く

☐ step [17,772]：名 段階

☐ neural crest [1,859]：名 神経堤

☐ development [48,509]：名 発生，発症，開発

訳 Wnt シグナル伝達は，神経堤発生の早期の段階に関与していると知られている

069

This novel approach can be **used to** assess quantitatively the effects of therapeutic interventions for treating liver failure.

☐ novel [30,313]：形 新規の，新しい

☐ approach [21,009]：名 アプローチ，方法（類 method [32,975]，strategy [13,405]，tool [6,214]）

☐ used to ～ [26,204]：～するために使われる

☐ assess the effect of ～ [681]：～の効果を評価する

☐ quantitatively [1,716]：副 定量的に

☐ therapeutic intervention [786]：治療介入

☐ treat [29,577]：動 治療する，処理する

☐ **liver failure** [239]：肝不全

訳 この新規のアプローチは，肝不全を治療することに対する治療介入の効果を定量的に評価するために使われうる

070 The present study was **designed to** evaluate this hypothesis.

☐ **present** [41,922]：形 現在の，存在する / 動 提示する

☐ **design** [14,086]：動 計画する，設計する（類 plan [1,911]），名 設計，計画

design to ～ [3,933]：～するために計画される

☐ **evaluate** [19,786]：動 評価する（類 assess [20,743]，estimate [13,499]）

☐ **hypothesis** [15,247]：名 仮説（類 assumption [1,542]，notion [1,569]，idea [2,167]，concept [2,660]）

訳 現在の研究は，この仮説を評価するために計画された

071 This study was **undertaken to** establish a rat model of chronic rejection.

☐ **undertake** [2,347]：動 着手する，計画する（類 attempt [3,991]，address [6,396]）

undertaken to ～ [1,050]：～するために着手される

☐ **establish** [19,016]：動 確立する（類 confirm [17,664]）

☐ **rat model** [646]：ラットモデル

☐ **chronic rejection** [575]：慢性拒絶反応

訳 この研究は，慢性拒絶反応のラットモデルを確立するために着手された

C. it を形式主語とする that 節

文例No			用例数
072	it … accepted that ～	～ということが受け入れられている	167
073	it … assumed that ～	～ということが推定される	329

072 **It is widely accepted that** tumors are monoclonal in origin.

- ☐ **widely** [5,189]：副 広く（類 broadly [1,060]，extensively [1,944]，universally [344]）
- ☐ **accept** [1,866]：動 受け入れる，認める（類 appreciate [452]）
  - it … accepted that ～ [167]：～ということが受け入れられている
- ☐ **tumor** [61,098]：名 腫瘍
- ☐ **monoclonal** [5,990]：形 単クローンの
- ☐ **origin** [7,942]：名 起源，由来

訳 腫瘍が単クローンの由来であるということは広く受け入れられている

073 **It is generally assumed that** protein kinase C is the sole receptor for phorbol esters in this system.

- ☐ **generally** [5,546]：副 一般に（類 commonly [4,265]，in general [1,797]）
- ☐ **assume** [2,937]：動 推定する（類 presume [1,195]），仮定する（類 hypothesize [6,370]）
  - it … assumed that ～ [329]：～ということが推定される
- ☐ **protein kinase** [16,532]：プロテインキナーゼ
- ☐ **sole** [1,014]：形 唯一の
- ☐ **receptor for ～** [3,144]：～に対する受容体
- ☐ **phorbol ester** [1,243]：ホルボールエステル
- ☐ **system** [54,451]：名 システム，系

訳 プロテインキナーゼCはこのシステムにおけるホルボールエステルに対する唯一の受容体であると一般に推定されている

3 自動詞＋前置詞の文例

自動詞は後ろに補語を伴うものと伴わないものとがあるが，補語を伴わない自動詞は前置詞を伴うことがしばしばある．このとき使われる前置詞には，個々の自動詞ごとに決まったパターンがあるので**自動詞＋前置詞**を習得しておくとよい．

A. 自動詞＋前置詞

文例No			用例数
074	protect against ～	～から保護する	1,800
075	serve as ～	～として働く，～として役立つ	7,135
076	act as ～	～として働く，～として作用する	6,441
077	result from ～	～に起因する	10,557
078	arise from ～	～から生じる	3,170
079	differ from ～	～と異なる	2,685
080	originate from ～	～から生じる	1,236
031	suffer from ～	～を患う	517
081	result in ～	～という結果になる	48,455
–	occur in ～	～において起こる	12,429
082	participate in ～	～参加する，～に関与する	5,828
083	exist in ～	～に存在する	2,475
084	differentiate into ～	～に分化する	1,111
085	consist of ～	～からなる	9,589
086	depend on ～	～に依存する	9,937
087	focus on ～	～に焦点を合わせる	4,044
088	rely on ～	～に頼る	1,668
041	contribute to ～	～に寄与する	19,994
089	lead to ～	～につながる	31,651
090	bind to ～	～に結合する	33,688
091	respond to ～	～に応答する，～に反応する	5,457

文例No			用例数
092	fail to *do*	～できない，～するのに失敗する	10,492
093	interact with～	～と相互作用する	14,220
094	interfere with～	～に干渉する	3,455
095	react with～	～と反応する	2,380

074 There are currently no vaccines to **protect against** a number of serious viral diseases such as SARS and AIDS.

□ **currently** [4,311]：副 現在（類 presently [557]，now [10,465]）

□ **vaccine** [10,085]：名 ワクチン

□ **protect** [9,261]：動 防ぐ，保護する（類 prevent [19,897]）

　protect against ～ [1,800]：～から保護する

□ **a number of ～** [7,385]：いくつかの～（類 several [34,093]）

□ **serious** [1,674]：形 重篤な（類 severe [12,151]，intense [1,624]）

□ **viral** [21,817]：形 ウイルスの，ウイルス性の

□ **disease** [71,437]：名 疾患

□ **such as ～** [23,848]：～のような

□ **SARS** [571]：名 重症急性呼吸器症候群（severe acute respiratory syndrome）

□ **AIDS** [2,994]：名 エイズ，後天性免疫不全症候群（acquired immunodeficiency syndrome）

訳 重症急性呼吸器症候群やエイズのようないくつかの重篤なウイルス性疾患から保護するワクチンは現在ない

075 Mites are known to **serve as** a vector and reservoir of the etiologic agent.

□ **mite** [247]：名 ダニ

□ **known to ～** [8,375]：～すると知られている

□ **serve as ～** [7,135]：～として働く，～として役立つ

□ **vector** [12,677]：名 媒介物，ベクター

□ **reservoir** [984]：名 宿主，貯蔵所

☐ **etiologic agent** [253]：病原体（類 pathogen [7,954]）

訳 ダニは，病原体の媒介物および宿主として働くことが知られている

076

Environmental agents producing point mutations may **act as** chromosome-breaking agents.

☐ **environmental** [5,140]：形 環境の

☐ **agent** [18,880]：名 作用物質，薬剤

☐ **produce** [38,705]：動 引き起こす（類 cause [46,697]，result in [48,455]，lead to [31,651]），産生する

☐ **point mutation** [2,795]：点突然変異

☐ **act** [19,853]：動 作用する，働く（類 serve [9,796]，function [95,000]，behave [1,367]，work [12,304]）

act as ～ [6,441]：～として働く，～として作用する

☐ **chromosome** [24,029]：名 染色体

☐ **break** [4,214]：動 切断する

訳 点突然変異を引き起こす環境作用物質は，染色体を切断する物質として作用するかもしれない

077

Structural abnormalities **resulting from** errors in embryogenesis or the fetal period are called congenital anomalies.

☐ **structural** [21,284]：形 構造的な

☐ **abnormality** [7,645]：名 異常（類 anomaly [914]，aberration [620]，defect [18,182]）

☐ **result from ～** [10,557]：～に起因する（類 derived from [11,249]，attributed to [2,837]）

☐ **error** [4,709]：名 エラー，誤り

☐ **embryogenesis** [1,811]：名 胚発生

☐ **fetal** [9,274]：形 胎児性の

☐ **period** [14,240]：名 期間（類 term [18,638]）

☐ **call** [6,053]：動（A を B と）呼ぶ（類 designate [4,156]，name [3,266]）

called [5,012]：～と呼ばれる

☐ **congenital** [1,980]：形 先天性の（類 inborn [98]）

□ **anomaly** [914]：名 異常（類 abnormality [7,645]）

訳 胚発生あるいは胎児期におけるエラーに起因する構造的な異常は，先天性異常と呼ばれる

078 Malignant lung tumors commonly **arise from** the respiratory epithelium.

□ **malignant** [4,511]：形 悪性の

□ **tumor** [61,098]：名 腫瘍

□ **commonly** [4,265]：副 一般に（類 generally [5,546]，usually [3,166]，normally [6,493]）

□ **arise from ～** [3,170]：～から生じる（類 originate from [1,236]，stem from [302]，derived from [11,249]）

□ **respiratory** [6,788]：形 呼吸の，呼吸器の

□ **epithelium** [7,400]：名 上皮

訳 悪性の肺腫瘍は，一般に呼吸上皮から生じる

079 Bacteria **differ from** fungi in the lack of ability to reproduce sexually or asexually.

□ **bacterium** [11,218]：名（複 bacteria）細菌

□ **differ** [11,437]：動 異なる（類 vary [9,676]）

differ from ～ [2,686]：～と異なる

□ **fungus** [1,716]：名（複 fungi）真菌

□ **lack** [23,888]：名 欠如（類 depletion [6,011]，loss [28,751]，deprivation [1,977]，defect [18,182]，deficit [3,988]）/ 動 欠く

□ **ability** [29,720]：名 能力（類 capacity [8,285]，capability [1,731]，potential [37,200]）

□ **ability to ～** [15,068]：～する能力

□ **reproduce** [1,486]：動 生殖する，再現する

□ **sexually** [864]：副 性的に，有性的に

□ **asexually** [16]：副 無性的に，無性生殖的に

訳 細菌は，有性的あるいは無性的に生殖する能力の欠如という点で真菌と異なる

080 Headache may **originate from** damage to pain-sensitive pathways of the peripheral or central nervous system.

- □ **headache** [456]：名 頭痛
- □ **originate** [2,139]：動 生じる（類 occur [42,905]）
  - **originate from ～** [1,236]：～から生じる（類 arise from [3,170], stem from [302], derived from [11,249]）
- □ **damage** [15,696]：名 障害, 損傷（類 injury [15,721], impairment [3,806], disorder [16,726]）/ 動 損傷する
- □ **pain** [5,403]：名 痛み
- □ **sensitive** [19,737]：形 感受性の（類 susceptible [4,520]）
- □ **pathway** [59,666]：名 経路
- □ **peripheral** [12,496]：形 末梢の
- □ **central nervous system** [3,405]：中枢神経系

訳 頭痛は，末梢あるいは中枢神経系の痛み感受性経路に対する障害から生じるかもしれない

081 The combined treatment **resulted in** improved survival.

- □ **combined treatment** [231]：併用療法
- □ **result in ～** [48,455]：～という結果になる（類 lead to [31,651], cause [46,697], produce [38,705]）
- □ **improve** [16,767]：動 改善する（類 advance [8,632]）
- □ **survival** [28,139]：名 生存，生存率

訳 併用療法は，改善された生存率という結果になった

082 The BRCA genes are suspected to **participate in** the development of sporadic breast cancer.

- □ **suspect** [1,509]：動 疑う，思う（類 think [8,004]）
- □ **participate** [6,477]：動 関与する，参加する
  - **participate in ～** [5,828]：～に関与する, ～に参加する（類 engage in [568], involved in [25,574], implicated in [7,864]）
- □ **development** [48,509]：名 発症（類 morbidity [3,142], onset [9,566]），発生，開発

□ **sporadic** [1,847]：形 散発性の，孤発性の

□ **breast cancer** [8,621]：乳癌

訳 BRCA遺伝子は，散発性乳癌の発症に関与すると思われている

083

The objective of this study was to determine whether differences **exist in** the trabecular bone structure of the femur and tibia.

□ **objective** [3,041]：名 目的（類 object [2,181]，purpose [4,516]，aim [4,574]，goal [3,570]）/ 形 客観的な

□ **determine whether ～** [7,898]：～かどうかを決定する

□ **difference** [33,598]：名 違い（類 divergence [2,132]，disparity [956]，distinction [762]）

□ **exist** [10,359]：動 存在する

exist in ～ [2,475]：～に存在する

□ **trabecular bone** [112]：海綿骨

□ **structure** [64,485]：名 構造

□ **femur** [197]：名 大腿骨

□ **tibia** [127]：名 脛骨

訳 この研究の目的は，大腿骨と脛骨の海綿骨構造に違いが存在するかどうかを決定することであった

084

Embryonic vessels **differentiate into** arteries and veins.

□ **embryonic** [10,071]：形 胚性の

□ **vessel** [6,144]：名 血管

□ **differentiate** [7,522]：動 分化する，区別する

differentiate into ～ [1,111]：～に分化する

□ **artery** [10,288]：名 動脈

□ **vein** [3,162]：名 静脈

訳 胚性の血管は，動脈と静脈に分化する

085 G protein α subunits **consist of** two domains: a GTPase domain and a helical domain.

- ☐ **G protein** [8,418]：G タンパク質
- ☐ **subunit** [38,120]：名 サブユニット
- ☐ **consist of ～** [9,589]：～からなる
- ☐ **domain** [87,363]：名 ドメイン
- ☐ **GTPase** [4,537]：名 GTP 加水分解酵素
- ☐ **helical** [4,963]：形 ヘリックスの，らせん状の

訳 Gタンパク質αサブユニットは2つのドメイン，すなわちGTP加水分解酵素ドメインとヘリックスドメインからなる

086 **Depending on** the dose administered, cystic fibrosis mice exhibited significantly higher mortality rates, compared to wild-type mice.

- ☐ **depend on ～** [9,937]：～に依存する（類 rely on [1,668]）
- ☐ **dose** [29,485]：名 用量，投与量
- ☐ **administer** [6,478]：動 投与する（類 give [18,279]），投薬する
- ☐ **cystic fibrosis** [1,288]：嚢胞性線維症
- ☐ **mouse** [123,326]：名（複 mice）マウス
- ☐ **exhibit** [27,339]：動 示す（類 display [15,251]，indicate [73,855]，show [148,875]）
- ☐ **significantly** [48,939]：副 有意に，著しく
- ☐ **higher** [31,357]：形（high の比較級）より高い
- ☐ **mortality rate** [1,805]：死亡率
- ☐ **compared to ～** [10,768]：～と比較される，～と比較して
- ☐ **wild-type** [39,564]：野生型の

訳 投与された用量に依存して，嚢胞性線維症マウスは野生型マウスと比較して有意に高い死亡率を示した

087 This review will **focus on** tissue engineering as a promising approach for cartilage regeneration and repair.

- ☐ **review** [12,123]：名 総説，概説，動 概説する

- ☐ **focus** [6,892]：動 焦点を合わせる，集中する（類 concentrate [2,317]），名 焦点

  focus on ～ [4,073]：～に焦点を合わせる
- ☐ **tissue engineering** [177]：ティッシュエンジニアリング，組織工学
- ☐ **promising** [2,161]：形 有望な
- ☐ **approach for ～** [1,923]：～のためのアプローチ
- ☐ **cartilage** [2,352]：名 軟骨
- ☐ **regeneration** [2,679]：名 再生
- ☐ **repair** [13,402]：名 修復（類 restoration [1,525]），動 修復する

訳 この総説は，軟骨再生および修復のための有望なアプローチとしてティッシュエンジニアリングに焦点を合わせるであろう

088 The test should not **rely on** the patient's ability to monitor or accurately record blood glucose levels.

- ☐ **rely on ～** [1,668]：～に頼る（類 depend on [9,937]）
- ☐ **ability** [29,720]：名 能力
- ☐ **monitor** [8,779]：動 モニターする（類 observe [51,935]），名 モニター
- ☐ **accurately** [1,951]：副 正確に（類 correctly [1,372]，precisely [1,175]，exactly [383]）
- ☐ **record** [8,607]：動 記録する / 名 記録
- ☐ **blood glucose level** [177]：血糖値

訳 その検査は血糖値をモニターし，あるいは正確に記録する患者の能力に頼るべきではない

089 The downstream events **leading to** retinal damage are poorly understood.

- ☐ **downstream** [9,565]：形 下流の
- ☐ **event** [24,512]：名 現象，事象
- ☐ **lead to ～** [31,651]：～につながる（類 result in [48,455]，contribute to [19,994]，cause [46,697]）
- ☐ **retinal damage** [41]：網膜障害
- ☐ **poorly understood** [3,015]：あまり理解されていない

訳 網膜障害につながる下流の現象は，あまり理解されていない

090 Despite decades of investigation, how insulin **binds to** its receptor remains largely unknown.

- □ **decade** [2,311]：名 十年
- □ **investigation** [6,218]：名 研究（類 study [167,914], research [8,330]), 調査（類 examination [6,511], test [45,529]）
- □ **insulin** [20,353]：名 インスリン
- □ **bind to ～** [33,688]：～に結合する
- □ **receptor** [125,789]：名 受容体
- □ **remain** [26,736]：動 ～のままである
- □ **largely** [6,410]：副 大部分は，（否定文で）ほとんど（類 mostly [1,734], predominantly [5,592], mainly [3,316]）
- □ **unknown** [11,802]：形 知られていない，未知の（類 unidentified [1,245]）

訳 数十年の研究にもかかわらず，どのようにインスリンがその受容体に結合するかはほとんど知られていないままである

091 Human memory T cells **respond to** microvascular endothelial cells and can injure allografts in vivo without priming.

- □ **human** [101,441]：形 ヒトの
- □ **memory T cell** [808]：メモリーT 細胞
- □ **respond to ～** [5,457]：～に応答する，～に反応する
- □ **microvascular endothelial cell** [479]：微小血管内皮細胞
- □ **injure** [1,498]：動 損傷する，傷害を与える（類 damage [15,696], impair [10,580]）
- □ **allograft** [5,871]：名 同種移植
- □ **in vivo** [41,680]：副 生体内で，形 生体内の
- □ **priming** [2,122]：名 初回刺激，プライミング

訳 ヒトのメモリーT 細胞は微小血管内皮細胞に応答し，そして生体内で初回刺激なしに同種移植を傷害しうる

092 A number of fertilized oocytes **failed to** develop beyond the two-cell stage.

- □ **a number of 〜** [7,385]：いくつかの〜
- □ **fertilize** [277]：動 受精させる
- □ **oocyte** [4,969]：名 卵母細胞
- □ **fail** [12,094]：動 〜できない，〜しない，失敗する
  - **fail to 〜** [10,492]：〜できない，〜するのに失敗する
- □ **develop** [38,325]：動 発達する，発生する，発症する，開発する
- □ **stage** [19,952]：名 期，ステージ，動 段階に分ける

訳 いくつかの受精した卵母細胞は，発生の２細胞期を越えて発達することができなかった

093 An intracellular protein known as Calmodulin **interacts with** a number of proteins involved in signal transduction.

- □ **intracellular** [20,566]：形 細胞内の
- □ **known as 〜** [3,485]：〜として知られている
- □ **Calmodulin** [3,539]：名 カルモジュリン
- □ **interact** [21,473]：動 相互作用する
  - **interact with 〜** [14,220]：〜と相互作用する
- □ **a number of 〜** [7,385]：いくつかの〜
- □ **involved in 〜** [25,574]：〜に関与する
- □ **signal transduction** [7,006]：シグナル伝達

訳 カルモジュリンとして知られている細胞内タンパク質は，いくつかのシグナル伝達に関与するタンパク質と相互作用する

094 The suppressor P25 may **interfere with** either assembly or function of the effector complexes of RNA silencing.

- □ **suppressor** [7,601]：名 サプレッサー，抑制因子
- □ **interfere** [5,056]：動 干渉する（類 interrupt [879]，impede [612]，hamper [615]）

interfere with ～ [3,455]：～に干渉する

□ **assembly** [14,093]：名 構築（類 formation [41,324]，construction [1,479]），集合

□ **function** [92,343]：名 機能，動 機能する

□ **effector** [8,929]：名 エフェクター，効果器

□ **complex** [80,256]：名 複合体，動 複合体を形成する，形 複雑な

□ **RNA silencing** [162]：RNA サイレンシング

訳 サプレッサーP25は，RNA サイレンシングのエフェクター複合体の構築あるいは機能のどちらかに干渉するかもしれない

095 We sought to prepare a T cell population capable of **reacting with** most adenoviruses that cause disease in immunocompromised patients.

□ **sought to ～** [2,674]：～しようと努めた

□ **prepare** [6,376]：動 調製する

□ **population** [28,214]：名 集団，人口

□ **capable of ～ ing** [7,651]：～できる

□ **react** [3,706]：動 反応する（類 respond [8,360]）

react with ～ [2,380]：～と反応する

□ **most** [44,482]：形 ほとんどの，副 最も

□ **adenovirus** [4,030]：名 アデノウイルス

□ **cause** [46,697]：動 引き起こす，名 原因

□ **disease** [71,437]：名 疾患

□ **immunocompromised patient** [161]：免疫不全患者

訳 我々は，免疫不全患者において疾患を引き起こすほとんどのアデノウイルスと反応できるT細胞集団を調製しようと努めた

II. 副詞の使い方

副詞は，主として文頭あるいは文中で用いられる．文全体を修飾する文頭の副詞はつなぎの表現としても用いられ，文章を構成するうえで非常に重要な役割を果たす．一方，文中の副詞は，しばしば形容詞，動詞（過去分詞）を前から修飾する．また，前置詞の前で動詞の強調として用いられることもある．このような副詞は，文にアクセントを付けるために必要な要素である．ここでは，1. **副詞＋過去分詞**，2. **副詞＋形容詞**，3. **副詞＋前置詞**，4. **文頭の副詞**に分類して示す．

1 副詞＋過去分詞の文例

文例No			用例数
096	highly conserved	高度に保存される	4,208
097	significantly reduced	有意に低下した	3,953
098	previously reported	以前に報告された	3,597
099	closely related	密接に関連した	3,216
089	poorly understood	あまり理解されていない	3,015
–	randomly assigned	無作為に割りあてられる	1,634
100	widely used	広く使われている	1,265
101	differentially expressed	差動的に発現される	1,243
–	recently identified	最近同定された	1,213
102	markedly reduced	顕著に低下した	1,152
–	commonly used	一般に使われる	1,112
103	stably transfected	安定的に遺伝子導入される	1,086
–	greatly reduced	大きく低下した	974
104	newly synthesized	新たに合成される	913
105	positively charged	正に荷電した	827
–	negatively charged	負に荷電した	820
106	completely blocked	完全にブロックされる	807

文例No			用例数
107	developmentally regulated	発生的に調節される	736
108	transiently transfected	一過性に遺伝子導入された	669
–	ubiquitously expressed	普遍的に発現される	652
–	structurally related	構造的に関連した	607
–	dramatically reduced	劇的に低下した	585
109	tightly regulated	堅固に調節される	493
110	genetically engineered	遺伝子改変される	448
111	partially purified	部分精製される	438
112	severely impaired	ひどく障害される	431
–	substantially reduced	実質的に低下した	423
113	not fully understood	完全には理解されていない	381
–	inversely related	逆に相関した	379
114	absolutely required	絶対的に必要とされる	339
115	extensively studied	広範に研究される	329
116	chronically infected	慢性的に感染した	324
–	specifically expressed	特異的に発現した	296
117	rapidly induced	急速に誘導される	270
118	experimentally determined	実験的に決定される	238
119	clearly defined	はっきりと明らかにされる	233

096 These mutations were distributed in **highly conserved** residues and were absent in 300 control chromosomes from an ethnically similar population.

□ **mutation** [70,231]：名 変異

□ **distribute** [3,661]：動 分布する（類 localize [15,589], locate [12,854], position [22,156]）

□ **conserve** [20,784]：動 保存する（類 preserve [2,565]）

highly conserved [4,208]：高度に保存される

□ **residue** [45,837]：名 残基

□ **absent** [4,385]：形 存在しない（類 deficient [18,199], defective [7,616] / 反 present [41,922]）

□ **control** [78,085]：名 対照群，コントロール，制御 / 動 制御する

□ **chromosome** [24,029]：名 染色体

□ **ethnically** [102]：副 民族的に

□ **similar** [45,428]：形 類似の，似ている

□ **population** [28,214]：名 集団，人口

訳 これらの変異は高度に保存された残基に分布しており，民族的に類似の集団の300の対照群の染色体には存在しなかった

097 The prevalence of diarrheal diseases should be **significantly reduced** in developing countries through improvements in hygiene to limit fecal-oral spread of enteric pathogens.

□ **prevalence** [5,415]：名 罹患率，流行（類 morbidity [3,142], susceptibility [7,728], onset [9,566]）

□ **diarrheal** [197]：形 下痢性の

□ **disease** [71,437]：名 疾患

□ **significantly reduced** [3,953]：有意に低下した

□ **developing country** [273]：発展途上国

□ **improvement** [6,188]：名 改善（類 refinement [787]）

□ **hygiene** [165]：名 衛生

□ **limit** [21,272]：動 制限する / 名 限界

□ **fecal** [768]：形 糞便の

- ☐ **oral** [6,343]：形 経口の，口腔の
- ☐ **spread** [4,686]：名 拡散，伝播（類 propagation [1,428]，prevalence [5,415]）／動 伝播する（類 propagate [1,246]）
- ☐ **enteric** [1,138]：形 腸内の，腸管の（類 enteral [289]）
- ☐ **pathogen** [7,954]：名 病原体（類 agent [18,880]）

訳 下痢性の疾患の流行は，腸内の病原体の糞便から口への拡散を制限する衛生の改善によって発展途上国において有意に低下するはずだ

098 A serine/threonine kinase expressed in the heart has been **previously reported** to regulate sodium channels.

- ☐ **serine** [9,365]：名 セリン
- ☐ **threonine** [3,492]：名 スレオニン
- ☐ **serine/threonine kinase** [953]：セリンスレオニンキナーゼ
- ☐ **expressed in ～** [18,536]：～において発現される
- ☐ **heart** [20,807]：名 心臓
- ☐ **previously reported** [3,597]：以前に報告された
- ☐ **regulate** [57,131]：動 調節する（類 modulate [13,347]，control [78,085]，mediate [71,309]）
- ☐ **sodium** [6,328]：名 ナトリウム（類 Na [10,284]）
- ☐ **channel** [33,765]：名 チャネル

訳 心臓において発現するセリンスレオニンキナーゼは，ナトリウムチャネルを調節することが以前に報告されている

099 It remains unresolved whether arthropods are more **closely related** to nematodes or to deuterostomes.

- ☐ **remain** [26,736]：動 ～のままである
- ☐ **unresolved** [532]：形 未解決の（類 unanswered [167]，unsolved [82]）
- ☐ **arthropod** [470]：名 節足動物
- ☐ **closely** [6,800]：副 密接に（類 tightly [2,848]，near [9,852]）
  closely related [3,216]：密接に関連した
- ☐ **nematode** [1,440]：名 線虫
- ☐ **deuterostome** [70]：名 後口動物

訳 節足動物が線虫により密接に関連するのか後口動物により密接に関連するのかは，未解決のままである

100 In some countries, a live anthrax vaccine has been **widely used** for prophylaxis against anthrax in both humans and animals.

- ☐ **country** [1,519]：名 国
- ☐ **live** [8,229]：形 生きた / 動 生きる
- ☐ **anthrax** [502]：名 炭疽菌
- ☐ **vaccine** [10,085]：名 ワクチン
- ☐ **widely used** [1,265]：広く使われている
- ☐ **used for ～** [3,818]：～のために使われる
- ☐ **prophylaxis** [1,038]：名 予防（類 prevention [3,977]）
- ☐ **humans** [9,989]：名（複数形）ヒト（human は，通常，形容詞として用いられる）

訳 いくつかの国において，生きた炭疽菌ワクチンは人と動物の両方において炭疽菌に対する予防のために広く使われてきた

101 The aim of this study was to identify **differentially expressed** genes in rat fetal liver epithelial stem cells during their proliferation, lineage commitment, and differentiation.

- ☐ **aim** [4,574]：名 目的 / 動 目的とする
- ☐ **identify** [73,456]：動 同定する
- ☐ **differentially** [4,015]：副 差動的に（類 differently [946]）
  differentially expressed [1,243]：差動的に発現される
- ☐ **rat** [41,664]：名 ラット
- ☐ **fetal** [9,274]：形 胎児性の
- ☐ **epithelial** [16,204]：形 上皮の
- ☐ **stem cell** [7,364]：幹細胞
- ☐ **proliferation** [19,120]：名 増殖（類 growth [59,775]，replication [21,221]）
- ☐ **lineage** [7,386]：名 系列
- ☐ **commitment** [1,067]：名 傾倒，コミットメント，運命付け

lineage commitment ［255］：分子系列決定

訳 この研究の目的は，それらの増殖，分子系列決定および分化の間にラットの胎肝上皮幹細胞において差動的に発現される遺伝子を同定することであった

102 The levels of Sp1 protein and mRNA were **markedly reduced** at postnatal day 15.

- □ level ［104,490］：名 レベル
- □ markedly reduced ［1,152］：顕著に低下した
- □ postnatal ［3,191］：形 出生後の

訳 Sp1タンパク質およびメッセンジャーRNAのレベルは，出生後15日目で顕著に低下した

103 In this study, we used estrogen-dependent human breast cancer cells **stably transfected** with the aromatase gene.

- □ use ［182,256］：動 使う / 名 使用
- □ estrogen ［6,639］：名 エストロゲン，女性ホルモン
- □ dependent ［70,633］：形 依存性の
- □ breast cancer ［8,621］：乳癌
- □ stably ［3,135］：副 安定的に

  stably transfected ［1,086］：安定的に遺伝子導入される
- □ transfect ［9,842］：動 遺伝子導入する，移入する（類 transform ［9,714］，transfer ［20,058］）
- □ aromatase ［358］：名 アロマターゼ

訳 この研究において，我々はアロマターゼ遺伝子を安定的に遺伝子導入されたエストロゲン依存性ヒト乳癌細胞を用いた

104 Secretion of **newly synthesized** proteins across the mammalian rough endoplasmic reticulum is known to be supported by two membrane proteins.

- □ secretion ［12,758］：名 分泌，分泌物
- □ newly synthesized ［913］：新たに合成される

□ **mammalian** [15,396]：形 哺乳類の

□ **rough endoplasmic reticulum** [92]：粗面小胞体

□ **known to ～** [8,375]：～すると知られている

□ **support** [28,621]：動 支持する，補助する（類 assist [3,032]）/ 名 支持

□ **membrane protein** [4,843]：膜タンパク質

訳 新たに合成されたタンパク質の哺乳類の粗面小胞体を越える分泌は２つの膜タンパク質によって補助されることが知られている

105

These results raise the possibility that a **positively charged** loop 2 is important to maintain processivity near physiologic ionic strength.

□ **raise the possibility that ～** [1,556]：～という可能性を示唆する

□ **positively charged** [827]：正に荷電した

□ **loop** [15,953]：名 ループ / 動 ループを作る

□ **important** [45,928]：形 重要な

□ **maintain** [13,333]：動 維持する（類 sustain [6,146]，keep [1,062]，hold [2,518]，retain [6,883]）

□ **processivity** [637]：名 処理能力

□ **physiologic** [2,327]：形 生理的な，生理学的な（類 physiological [9,754]）

□ **ionic strength** [1,104]：イオン強度

訳 これらの結果は，正に荷電したループ２が生理的なイオン強度に近い処理能力を維持するために重要であるという可能性を示唆している

106

In lean mice, food intake was almost **completely blocked** by the fatty acid synthase inhibitor, C75.

□ **lean** [986]：形 痩せた

□ **food intake** [933]：食物摂取

□ **almost** [4,947]：副 ほとんど（類 nearly [5,643]，virtually [2,089]，mostly [1,734]，largely [6,410]）

□ **completely** [8,911]：副 完全に（類 entirely [1,497]，fully [7,013]，sufficiently [944]，enough [1,125]）

completely blocked [807]：完全にブロックされる

□ **fatty acid synthase** [274]：脂肪酸合成酵素

☐ **inhibitor** [43,673]：名 阻害剤

訳 痩せたマウスにおいて，食物摂取は脂肪酸合成酵素阻害剤C75によってほとんど完全にブロックされた

107

The border cells in the ovary undergo a well-defined and **developmentally regulated** cell migration.

☐ **border** [1,538]：名 境界，辺縁

☐ **ovary** [2,381]：名 卵巣

☐ **undergo** [19,420]：動 起こす，経験する（類 experience [6,331]）

☐ **well-defined** [1,616]：形 詳細に明らかにされた

☐ **developmentally** [1,210]：副 発生的に，発達的に

developmentally regulated [736]：発生的に調節される

☐ **regulate** [57,131]：動 調節する

☐ **migration** [10,131]：名 移動，遊走（類 movement [8,094]，transfer [20,058]）

訳 卵巣における境界細胞は詳細に明らかにされ，発生的に調節された細胞移動を起こす

108

Upstream stimulatory factor seems to be necessary for full promoter activity in **transiently transfected** cells.

☐ **upstream stimulatory factor** [83]：上流刺激因子（USF）

☐ **seem to ～** [3,153]：～するように思われる

☐ **necessary for ～** [6,750]：～のために必要な

☐ **full** [12,117]：形 完全な（類 complete [12,150]，perfect [519]）

☐ **promoter activity** [3,379]：プロモーター活性

☐ **transiently** [2,921]：副 一過性に

transiently transfected [669]：一過性に遺伝子導入された

訳 上流刺激因子は，一過性に遺伝子導入された細胞において完全なプロモーター活性のために必要であるように思われる

109 These survival and apoptotic signals are **tightly regulated** by a large number of molecules.

- ☐ **survival** [28,139]：名 生存，生存率
- ☐ **apoptotic** [8,348]：形 アポトーシスの，アポトーシス性の
- ☐ **signal** [40,955]：名 シグナル，信号 / 動 信号を送る
- ☐ **tightly** [2,848]：副 堅固に，密接に（類 closely [6,800]）
  tightly regulated [493]：堅固に調節される
- ☐ **a large number of ～** [1,348]：多数の～（類 many [26,628]，numerous [4,565]，a host of [125]）
- ☐ **molecule** [36,845]：名 分子

訳 これらの生存とアポトーシスの信号は多数の分子によって堅固に調節される

110 Improved recombinant DNA technology is expected to lead to development of **genetically engineered** inhibitors, such as dominant-negative mutants.

- ☐ **improve** [16,767]：動 改善する
- ☐ **recombinant** [17,251]：形 組換え型の / 名 組換え体
- ☐ **technology** [4,040]：名 技術，技法，テクノロジー（類 technique [14,941]）
- ☐ **expected to ～** [1,448]：～すると予想される，～すると期待される
- ☐ **lead to ～** [31,651]：～につながる
- ☐ **development** [48,509]：名 開発，発生，発症
- ☐ **genetically engineered** [448]：遺伝子改変される
- ☐ **inhibitor** [43,673]：名 阻害物質，阻害剤
- ☐ **dominant-negative** [5,875]：形 ドミナントネガティブな，優性阻害の
- ☐ **mutant** [79,727]：名 変異体 / 形 変異の

訳 改良された組換えDNA技術は，ドミナントネガティブ変異体のような遺伝子改変された阻害物質の開発につながることが期待される

111

Replacement therapy using **partially purified** human enzyme has proved to be biochemically effective in short-term pilot trials.

- □ **replacement therapy** [614]：補充療法
- □ **partially** [8,729]：副 部分的に（類 partly [1,346]，in part [6,718]）
  - **partially purified** [438]：部分精製される
- □ **purify** [15,730]：動 精製する（類 isolate [34,355]）
- □ **enzyme** [49,164]：名 酵素
- □ **prove** [4,555]：動 判明する，証明する（類 evidence [35,499]，demonstrate [80,078]）
- □ **biochemically** [682]：副 生化学的に
- □ **effective** [14,834]：形 効果的な
- □ **short-term** [2,459]：短期間の
- □ **pilot** [497]：形 試験的な
- □ **trial** [13,519]：名 治験，試行

訳 部分精製されたヒト酵素を使った補充療法は，短期間の試験的な治験において生化学的に効果的であることが判明している

112

Cognitive development is **severely impaired** in individuals with Williams syndrome.

- □ **cognitive development** [38]：認知発達
- □ **severely** [2,928]：副 ひどく，激しく（類 markedly [7,503]，significantly [48,939]，dramatically [4,756]，drastically [559]，strongly [11,204]）
  - **severely impaired** [431]：ひどく障害される
- □ **individual** [24,178]：名 個々人（類 person [5,218]）/ 形 個々の
- □ **Williams syndrome** [43]：ウィリアムズ症候群

訳 認知発達は，ウィリアムズ症候群の個々人においてひどく障害されている

113

The mechanisms that cause these translocations are **not fully understood**.

- □ **cause** [46,697]：動 引き起こす / 名 原因
- □ **translocation** [9,431]：名 転位置，転座，移行（類 shift [12,913]，

transition [12,200], transfer [20,058], transposition [1,168])

☐ **fully** [7,013]：副 完全に（類 entirely [1,497], completely [8,911], sufficiently [944], enough [1,125]）

not fully understood [381]：完全には理解されていない

訳 これらの転位置を起こす機構は完全には理解されていない

114 A nucleotide containing a phosphate group complexed with magnesium ion appears to be **absolutely required** for binding.

☐ **nucleotide** [19,659]：名 ヌクレオチド

☐ **contain** [71,403]：動 含む

☐ **phosphate** [10,119]：名 リン酸

☐ **group** [69,369]：名 基，群 / 動 グループ化する

☐ **complex** [80,256]：動 複合体を形成する / 名 複合体 / 形 複雑な

complexed with ～ [1,033]：～と複合体を形成した

☐ **magnesium** [1,128]：名 マグネシウム

☐ **ion** [15,384]：名 イオン

☐ **appear to ～** [21,779]：～するように思われる

☐ **absolutely** [692]：副 絶対的に，全く（類 completely [8,911], entirely [1,497], fully [7,013], quite [1,408], thoroughly [241], totally [605]）

absolutely required [339]：絶対的に必要とされる

☐ **binding** [133,851]：名 結合

訳 マグネシウムイオンと複合体を形成するリン酸基を含むヌクレオチドは，結合のために絶対的に必要とされるようである

115 The role of this regulatory factor in vertebrate myogenesis has been **extensively studied**.

☐ **role** [87,150]：名 役割

☐ **regulatory** [19,642]：形 制御の，調節性の

☐ **factor** [103,476]：名 因子

☐ **vertebrate** [6,044]：名 脊椎動物

☐ **myogenesis** [457]：名 筋形成

- □ **extensively** [1,944]：副 広範に, 大規模に（類 widely [5,189], broadly [1,060], universally [344], ubiquitously [914]）
  extensively studied [329]：広範に研究される

訳 脊椎動物の筋形成におけるこの制御因子の役割は広範に研究されてきた

116

Patients with defects in cellular immune competence are more likely to remain **chronically infected** rather than to clear the virus.

- □ **patient with ～** [43,273]：～の患者
- □ **defect** [18,182]：名 欠損（類 deficiency [6,979], deficit [3,988]）
- □ **cellular** [29,968]：形 細胞の, 細胞性の
- □ **immune** [20,249]：形 免疫の
- □ **competence** [1,025]：名 能力
- □ **more likely to ～** [2,144]：おそらくより～しそうである
- □ **remain** [26,736]：動 ～のままである
- □ **chronically** [1,180]：副 慢性的に（反 acutely [1,120]）
  chronically infected [324]：慢性的に感染した
- □ **rather than ～** [7,606]：～よりむしろ
- □ **clear** [5,786]：動 除去する / 形 明らかな
- □ **virus** [47,464]：名 ウイルス

訳 細胞性免疫能の欠損を持つ患者は, そのウイルスを除去するよりもむしろおそらく慢性的に感染したままでありそうである

117

We observed **rapidly induced** changes in a cortically mediated perception in human subjects.

- □ **observe** [51,935]：動 観察する（類 monitor [8,779]）, 認める（類 identify [73,456]）
- □ **rapidly induced** [270]：急速に誘導される
- □ **change in ～** [30,981]：～の変化
- □ **cortically** [46]：副 皮質性に
- □ **mediated** [52,559]：仲介される
- □ **perception** [1,594]：名 知覚, 認知

☐ subject [26,037]：名 対象者 / 動 受けさせる

訳 我々は，ヒトの対象者において皮質性に仲介された認知の急速に誘導される変化を観察した

118 We performed a comparative analysis of all protein-antibody complexes for which structures have been **experimentally determined**.

☐ perform [21,488]：動 行う，実行する（類 conduct [6,182]，carry out [3,749]，undertake [2,347]）

☐ comparative [2,706]：形 比較による

☐ analysis [85,671]：名 解析，分析

☐ antibody [36,724]：名 抗体

☐ complex [80,256]：名 複合体 / 動 複合体を形成する / 形 複雑な

☐ structure [64,485]：名 構造

☐ experimentally [2,409]：副 実験的に

experimentally determined [238]：実験的に決定される

訳 我々は，その構造が実験的に決定されたすべてのタンパク質－抗体複合体の比較分析を実行した

119 The mechanisms responsible for bone resorption inhibition have not been **clearly defined**.

☐ responsible for ～ [12,456]：～に対して責任のある，～の原因である

☐ resorption [850]：名 再吸収，吸収

☐ inhibition [37,129]：名 抑制（類 suppression [7,483]，depression [5,911]，repression [6,339]）

☐ clearly defined [233]：はっきりと明らかにされる

訳 骨の再吸収抑制の原因である機構は，はっきりとは明らかにされていない

2 副詞＋形容詞の文例

文例No			用例数
120	statistically significant	統計学的に有意な	2,605
121	constitutively active	構成的に活性のある	2,154
122	physiologically relevant	生理学的に関連する	680
160	commercially available	市販されている	622
246	currently available	現在利用できる	552
123	relatively little	比較的わずかしかない	461
124	potentially important	潜在的に重要な	432
125	much larger	ずっとより大きい	423
126	approximately equal	おおよそ等しい	360
127	mutually exclusive	相互に排他的な	355
128	remarkably similar	著しく類似している	299
129	distinctly different	明らかに異なる	268
130	slightly higher	わずかにより高い	266
131	primarily responsible for ～	～の主な原因である	225

120 There were no **statistically significant** differences in demographics between the experimental group and the control group.

- □ **statistically** [3,729]：副 統計学的に
  statistically significant [2,605]：統計学的に有意な
- □ **difference** [33,598]：名 違い
- □ **demographics** [340]：名 人口統計，人口統計学
- □ **experimental** [12,406]：形 実験の
- □ **group** [69,369]：名 群，グループ，基 / 動 グループ化する
- □ **control** [78,085]：名 対照群，コントロール，制御 / 動 制御する

訳 実験群と対照群の間に人口統計の統計学的に有意な差はなかった

121 Transgenic mice expressing **constitutively active** protein kinase C in the colon were found to be highly susceptible to carcinogen-induced colon carcinogenesis.

- □ **transgenic** [14,331]：形 トランスジェニック，遺伝子導入の
- □ **express** [78,977]：動 発現する
- □ **constitutively** [5,463]：副 構成的に，恒常的に
  - **constitutively active** [2,154]：構成的に活性のある
- □ **protein kinase** [16,532]：プロテインキナーゼ
- □ **colon** [4,699]：名 大腸，結腸
- □ **found to ～** [15,859]：～することが見つけられる
- □ **highly susceptible** [285]：高度に感受性の
- □ **carcinogen** [936]：名 発癌物質
- □ **～ -induced** [4,635]：～に誘導される
- □ **carcinogenesis** [1,739]：名 発癌

訳 大腸において構成的に活性のあるプロテインキナーゼCを発現するトランスジェニックマウスは，発癌物質誘導性の大腸発癌に高度に感受性であることが見つけられた

122 We sought to evaluate **physiologically relevant** interactions between ligands for this versatile transport protein.

- □ **sought to ～** [2,674]：～しようと努めた
- □ **evaluate** [19,786]：動 評価する
- □ **physiologically** [1,444]：副 生理学的に
  - **physiologically relevant** [680]：生理学的に関連する
- □ **interaction between** [11,513]：～の間の相互作用
- □ **ligand** [31,861]：名 リガンド
- □ **versatile** [524]：形 万能の，多目的の
- □ **transport** [18,476]：名 輸送（類 transfer [20,058]，delivery [5,914]）／動 輸送する

訳 我々は，この万能の輸送タンパク質に対するリガンドの間の生理学的に関連する相互作用を評価しようと努めた

123 **Relatively little** is known about the peripheral mechanisms that excitatory amino acid receptors may regulate when activated.

- ☐ **relatively little** [461]：比較的わずかしかない
- ☐ **know** [34,957]：動 知る（類 learn [1,427], understand [12,582]）
- ☐ **peripheral** [12,496]：形 末梢の
- ☐ **excitatory amino acid** [229]：興奮性アミノ酸
- ☐ **receptor** [125,789]：名 受容体
- ☐ **regulate** [57,131]：動 調節する
- ☐ **activate** [58,997]：動 活性化する（類 potentiate [2,958], enhance [32,256], augment [3,582]）

訳 興奮性アミノ酸受容体が活性化されたときに調節するかもしれない末梢の機構について比較的わずかしか知られていない

124 These novel approaches may serve as **potentially important** therapeutic interventions in the treatment of advanced colorectal cancer.

- ☐ **novel** [30,313]：形 新規の，新しい
- ☐ **approach** [21,009]：名 アプローチ，方法
- ☐ **serve as ～** [7,135]：～として役立つ，～として働く
- ☐ **potentially important** [432]：潜在的に重要な
- ☐ **therapeutic intervention** [786]：治療介入
- ☐ **treatment** [60,138]：名 治療（類 therapy [28,037], care [11,307], practice [3,583]），処理
- ☐ **advanced** [3,984]：進行した
- ☐ **colorectal cancer** [1,720]：結腸直腸癌

訳 これらの新規のアプローチは，進行した結腸直腸癌の治療における潜在的に重要な治療介入として役立つかもしれない

125 Smoking is a **much larger** risk factor for cardiovascular disease mortality than fine particulate matter air pollution.

- □ **smoking** [3,471]：名 喫煙
- □ **larger** [7,028]：形 より大きな
  much larger [423]：ずっとより大きい
- □ **risk factor** [6,751]：危険因子，リスク因子
- □ **cardiovascular disease** [1,853]：循環器疾患
- □ **mortality** [12,864]：名 死亡率（類 lethality [1,790]）
- □ **fine** [1,496]：形 微細な
- □ **particulate** [558]：形 粒子状の
- □ **matter** [2,581]：名 物質，問題（類 substance [2,497]，material [5,476]，problem [6,340]）
- □ **air** [2,571]：名 空気，大気
- □ **pollution** [212]：名 汚染（類 contamination [639]）

訳 喫煙は，循環器疾患による死亡率に対して微細な粒子状物質による大気汚染よりもずっと大きなリスク因子である

126 These two proteins were detected in **approximately equal** amounts in the cell.

- □ **detected in 〜** [9,116]：〜において検出される
- □ **equal** [2,730]：形 等しい，同じ（類 equivalent [4,744]，identical [9,552]，same [24,047]，consistent [24,134]，comparable [6,413]）
  approximately equal [360]：おおよそ等しい
- □ **amount** [11,659]：名 量（類 volume [11,239]，content [8,476]，quantity [2,013]）

訳 これら２つのタンパク質が，その細胞においておおよそ等しい量で検出された

127 These eight genes were found to be mutated in a **mutually exclusive** manner.

- □ **found to 〜** [15,859]：〜することが見つけられる

□ **mutate** [5,677]：動 変異させる

□ **mutually exclusive** [355]：相互に排他的な

□ **manner** [13,342]：名 様式

訳 これらの８つの遺伝子は，相互に排他的な様式で変異していることが見つけられた

128

The inner ear defects in these mice are **remarkably similar** to those seen in the corresponding human condition.

□ **inner ear** [638]：内耳

□ **defect** [18,182]：名 欠損

□ **remarkably** [2,458]：副 著しく（類 markedly [7,503]，strikingly [1,715]，dramatically [4,756]，notably [1,715]，extremely [2,414]，significantly [48,939]）

remarkably similar [299]：著しく類似している

□ **corresponding** [10,979]：形 対応する

□ **condition** [30,859]：名 状態，条件（類 state [36,521]，situation [1,437]）/ 動 条件づける

訳 これらのマウスにおける内耳の欠陥は，対応するヒトの状態において見られるそれらと著しく類似している

129

Distinctly different patterns of gene expression were observed at different levels of osteoprogenitor maturation.

□ **distinctly** [473]：副 明らかに（類 apparently [3,270]，clearly [3,471]，evidently [199]，obviously [110]，unambiguously [351]）

distinctly different [268]：明らかに異なる

□ **pattern** [33,282]：名 パターン，様式

□ **gene expression** [21,171]：遺伝子発現

□ **observe** [51,935]：動 観察する

□ **osteoprogenitor** [46]：名 骨前駆細胞

□ **maturation** [5,412]：名 成熟

訳 遺伝子発現の明らかに異なるパターンが，骨前駆細胞成熟の異なるレベルで観察された

130

In antibody testing, females had a **slightly higher** incidence of positivity than males.

□ **antibody** [36,724]：名 抗体

□ **testing** [6,093]：名 テスト，試験（類 examination [6,511]，test [45,529]，assessment [5,256]）

□ **female** [9,873]：名 女性，雌（類 woman [18,210]）

□ **slightly** [3,481]：副 わずかに（類 modestly [775]，marginally [437]，weakly [2,021]）

slightly higher [266]：わずかにより高い

□ **incidence** [8,627]：名 発生率，頻度

□ **positivity** [378]：名 陽性

□ **male** [12,104]：名 男性，雄（類 man [10,301]）

訳 抗体テストにおいて，女性は男性よりも陽性の頻度がわずかに高かった

131

The activated proteases are **primarily responsible for** the destruction of the bacteria.

□ **activate** [58,997]：動 活性化する

□ **protease** [10,907]：名 プロテアーゼ

□ **primarily responsible for ～** [225]：～の主な原因である

□ **destruction** [2,152]：名 破壊（類 disruption [6,656]，ablation [2,461]，lesion [18,776]）

□ **bacterium** [11,218]：名（複 bacteria）細菌

訳 活性化されたプロテアーゼが，細菌の破壊の主な原因である

3 副詞＋前置詞の文例

文例No			用例数
406	shortly after ～	～のあと直ちに	573
132	directly by ～	直接～によって	434
133	possibly by ～	もしかしたら～によって	414
134	presumably by ～	おそらく～によって	303
135	solely by ～	もっぱら～によって	250
136	perhaps by ～	おそらく～によって	217
137	primarily in ～	主に～において	1,036
138	exclusively in ～	～において独占的に	879
139	predominantly in ～	主に～において	911
–	specifically in ～	～において特異的に	833
140	mainly in ～	主に～において	448
141	specifically to ～	～に特異的に	1,248
142	preferentially to ～	～に優先的に	312
143	selectively to ～	～に選択的に	253
144	together with ～	～と合わせると，～とともに	4,371
145	linearly with ～	～とともに直線的に	295
146	simultaneously with ～	～と同時に	271

132 β-adrenergic blockers lower arterial pressure **directly by** reducing systemic vascular resistance.

□ **adrenergic blocker** [27]：アドレナリン遮断薬

□ **lower** [25,400]：動 低下させる（類 decrease [51,139], reduce [57,875], diminish [5,477], down-regulate [2,897]）, 形 より低い

□ **arterial pressure** [864]：動脈圧

□ **directly** [16,866]：副 直接（反 indirectly [1,618]）

directly by ～ [434]：直接～によって

□ **systemic** [7,731]：形 全身性の（反 local [9,593]）

□ **vascular resistance** [628]：血管抵抗

訳 βアドレナリン遮断薬は，全身性の血管抵抗を直接低下させることによって動脈圧を低下させる

133 Human herpesvirus 6 is an immunosuppressive and neurotropic virus that may be transmitted by saliva and **possibly by** genital secretions.

□ **human herpesvirus** [429]：ヒトヘルペスウイルス

□ **immunosuppressive** [1,647]：形 免疫抑制性の

□ **neurotropic** [137]：形 向神経性の

□ **virus** [47,464]：名 ウイルス

□ **transmit** [2,545]：動 伝染させる，伝える（類 transport [18,476], transfer [20,058], infect [22,659]）

□ **saliva** [642]：名 唾液

□ **possibly** [4,640]：副 もしかしたら（類 potentially [6,206], probably [5,067], presumably [2,634], perhaps [2,517]）

possibly by ～ [414]：もしかしたら～によって

□ **genital** [1,036]：形 生殖の（類 reproductive [2,076]），生殖器の

□ **secretion** [12,758]：名 分泌物，分泌

訳 ヒトヘルペスウイルス6は，唾液によって，そしてもしかしたら生殖器の分泌物によって伝搬されるかもしれない免疫抑制性で向神経性のウイルスである

134

Suramin actively treats prostate cancer, **presumably by** blocking the action of growth factors such as FGF.

- □ **suramin** [290]：名 スラミン
- □ **actively** [1,440]：副 能動的に，活発に（類 vigorously [142]，effectively [3,808]，positively [3,375]）
- □ **treat** [29,577]：動 治療する，処理する
- □ **prostate cancer** [4,808]：前立腺癌
- □ **presumably** [2,634]：副 おそらく（類 potentially [6,206]，probably [5,067]，possibly [4,640]，perhaps [2,517]）

  presumably by 〜 [303]：おそらく〜によって
- □ **block** [31,876]：動 ブロックする，遮断する（類 abrogate [3,332]，silence [4,716]，prevent [19,897]，suppress [12,609]）／名 遮断
- □ **action** [17,526]：名 作用
- □ **growth factor** [18,400]：増殖因子，成長因子
- □ **FGF** [3,891]：名 FGF，線維芽細胞増殖因子（fibroblast growth factor）

訳 スラミンは，おそらくはFGFのような成長因子の作用を遮断することによって，能動的に前立腺癌を治療する

135

These results indicate that the toxic effects of glutamate receptor overstimulation can be explained **solely by** calcium influx.

- □ **indicate that 〜** [58,147]：〜ということを示す
- □ **toxic** [2,871]：形 有毒な（類 virulent [1,322]）
- □ **effect** [106,593]：名 影響，効果（類 influence [12,979]，impact [6,448]，efficacy [8,112]）
- □ **glutamate receptor** [1,753]：グルタミン酸受容体
- □ **overstimulation** [35]：名 過剰刺激（類 hyperstimulation [17]）
- □ **explain** [9,556]：動 説明する
- □ **solely** [1,439]：副 もっぱら（類 exclusively [3,562]，only [56,980]），単に（類 simply [1,148]，merely [332]）

  solely by 〜 [250]：もっぱら〜によって
- □ **calcium influx** [454]：カルシウム流入

訳 これらの結果は，グルタミン酸受容体の過剰刺激の有毒な影響はもっぱらカルシウム流入によって説明されうるということを示す

136 The neuropeptide galanin gene expression in the cholinergic basal forebrain may be negatively regulated, **perhaps by** factors concomitant with puberty.

☐ **neuropeptide** [1,482]：名 神経ペプチド，ニューロペプチド

☐ **galanin** [408]：名 ガラニン

☐ **gene expression** [21,171]：遺伝子発現

☐ **cholinergic** [2,123]：形 コリン作動性の

☐ **basal forebrain** [399]：前脳基底核

☐ **negatively** [3,506]：副 負に（反 positively [3,375]）

☐ **regulate** [57,131]：動 調節する

☐ **perhaps** [2,517]：副 おそらく，多分（類 probably [5,067]，presumably [2,634]，possibly [4,640]）

perhaps by ～ [217]：おそらく～によって

☐ **factor** [103,476]：名 因子

☐ **concomitant with ～** [697]：～に随伴して

☐ **puberty** [227]：名 思春期

訳 コリン作動性前脳基底核における神経ペプチドガラニン遺伝子発現は，おそらく思春期に随伴する因子によって，負に調節されるかもしれない

137 ABCC6 is expressed **primarily in** the liver and to a lesser degree in the kidney.

☐ **express** [78,977]：動 発現する

☐ **primarily** [8,336]：副 主に（類 mainly [3,316]，mostly [1,734]，primarily [8,336]，principally [668]，predominantly [5,592]）

primarily in ～ [1,036]：主に～において

☐ **to a lesser degree** [240]：より少ない程度で

☐ **kidney** [10,654]：名 腎臓

訳 ABCC6は主に肝臓において，そしてより少ない程度で腎臓において発現する

138 This orphan receptor tyrosine kinase is expressed almost **exclusively in** endothelial cells.

- □ **orphan receptor** [208]：オーファン受容体
- □ **tyrosine kinase** [6,765]：チロシンキナーゼ
- □ **express** [78,977]：動 発現する
- □ **almost** [4,947]：副 ほとんど
- □ **exclusively** [3,562]：副 独占的に，もっぱら（類 exclusively [3,562]，only [56,980]），単に（類 simply [1,148]，merely [332]）

  exclusively in ～ [879]：～において独占的に
- □ **endothelial cell** [10,488]：内皮細胞

訳 このオーファン受容体チロシンキナーゼは，内皮細胞においてほとんど独占的に発現される

139 Lymphangioleiomyomatosis is the result of aberrant smooth muscle proliferation and occurs in a sporadic form **predominantly in** premenopausal women.

- □ **lymphangioleiomyomatosis** [30]：名 リンパ脈管筋腫症
- □ **aberrant** [2,859]：形 異常な（類 abnormal [6,231]，unusual [3,427]）
- □ **smooth muscle** [5,630]：平滑筋
- □ **proliferation** [19,120]：名 増殖
- □ **occur** [42,905]：動 起こる，生じる
- □ **sporadic** [1,847]：形 散発性の，孤発性の
- □ **form** [9,556]：名 型 / 動 形成する
- □ **predominantly** [5,592]：副 主に（類 mainly [3,316]，mostly [1,734]，primarily [8,336]，largely [6,410]）

  predominantly in ～ [911]：主に～において
- □ **premenopausal** [307]：形 閉経前の

訳 リンパ脈管筋腫症は，異常な平滑筋増殖の結果であり，主に閉経前の女性において散発性の型で起こる

140 Elemental mercury in blood is excreted **mainly in** the urine and feces.

- □ **elemental** [297]：形 基本的な，元素の
- □ **mercury** [442]：名 水銀
- □ **blood** [29,091]：名 血液
- □ **excrete** [290]：動 排泄する（類 eliminate [6,012]）
- □ **mainly** [3,316]：副 主に（類 predominantly [5,592]，mostly [1,734]，primarily [8,336]，largely [6,410]）

  mainly in ～ [448]：主に～において
- □ **urine** [2,266]：名 尿
- □ **feces** [218]：名 糞便（類 stool [640]）

訳 血液中の元素水銀は，主に尿と糞便に排泄される

141 This membrane-active antifungal agent was found to bind **specifically to** amyloid fibrils.

- □ **membrane** [56,085]：名 膜
- □ **active** [33,441]：形 活性の，活性のある
- □ **antifungal** [565]：形 抗真菌の
- □ **agent** [18,880]：名 薬剤，作用物質
- □ **found to ～** [15,859]：～することが見つけられる
- □ **bind** [187,860]：動 結合する（類 associate [82,102]，attach [3,455]，couple [17,384]，link [28,890]）
- □ **specifically** [14,032]：副 特異的に（類 particularly [7,227]，especially [4,892]）

  specifically to ～ [1,248]：～に特異的に
- □ **amyloid** [4,619]：名 アミロイド
- □ **fibril** [2,171]：名 原繊維

訳 この膜活性抗真菌薬は，アミロイド線維に特異的に結合することが見つけられた

142 Each of the recombinant proteins was found to bind **preferentially to** a specific fragment of minicircle DNA.

- □ **recombinant** [17,251]：名 組換え体 / 形 組換え型の
- □ **found to 〜** [15,859]：〜することが見つけられる
- □ **bind** [187,860]：動 結合する
- □ **preferentially** [4,208]：副 優先的に（類 predominantly [5,592]）
  preferentially to 〜 [312]：〜に優先的に
- □ **specific** [85,296]：形 特異的な
- □ **fragment** [15,575]：名 断片 / 動 断片化する
- □ **minicircle** [127]：形 小環状の / 名 小円

訳 組換え体タンパク質のおのおのは，小環状のDNAの特異的な断片に優先的に結合することが見つけられた

143 These proteins were engineered to respond rapidly and **selectively to** their target.

- □ **engineer** [4,216]：動 操作する（類 manipulate [1,329]，operate [3,663]），設計する（類 design [14,086]）/ 名 技師
- □ **respond** [8,360]：動 応答する，反応する（類 reacfvt [3,706]）
- □ **rapidly** [9,988]：副 急速に，迅速に（類 quickly [824]，briefly [430]）
- □ **selectively** [6,466]：副 選択的に
  selectively to 〜 [253]：〜に選択的に
- □ **target** [51,095]：名 標的 / 動 標的にする

訳 これらのタンパク質は，それらの標的に対して迅速にそして選択的に応答するように操作された

144 These results, **together with** previous reports, suggest that the expression of an unidentified DNA polymerase may account for the mutant phenotype.

- □ **together with 〜** [4,371]：〜と合わせると，〜とともに
- □ **previous** [16,852]：形 以前の（類 prior [8,802]，former [1,231]，recent [18,452]）
- □ **report** [54,972]：名 報告 / 動 報告する

☐ suggest that ～ [96,112]：～ということを示唆する

☐ expression [154,475]：発現

☐ unidentified [1,245]：形 未同定の，未知の（類 unknown [11,802]）

☐ polymerase [16,312]：名 ポリメラーゼ

☐ account for ～ [8,259]：～を説明する（類 explain [9,556]，illustrate [2,870]），～を占める

☐ mutant [79,727]：名 変異体 / 形 変異の

☐ phenotype [23,661]：名 表現型

訳 これらの結果は，以前の報告と合わせると，未同定のDNAポリメラーゼの発現がその変異体の表現型を説明するかもしれないことを示唆している

145 The seroprevalence of the human T-cell leukemia virus type 2 was found to increase **linearly with** age.

☐ seroprevalence [186]：名 血清有病率

☐ human T-cell leukemia virus [225]：ヒトT細胞白血病ウイルス

☐ found to ～ [15,859]：～することが見つけられる

☐ increase [146,670]：動 増大する，増大させる / 名 増大

☐ linearly [766]：副 直線的に

linearly with ～ [295]：～とともに直線的に

訳 ヒトT細胞白血病ウイルス2型の血清有病率は，年齢とともに直線的に増大することが見つけられた

146 These infections are clinically indistinguishable from Mycoplasma pneumoniae infection and may in fact occur **simultaneously with** mycoplasmal infections.

☐ infection [44,601]：名 感染

☐ clinically [4,700]：副 臨床的に

☐ indistinguishable from ～ [1,171]：～と区別できない

☐ Mycoplasma pneumoniae [67]：肺炎マイコプラズマ

☐ in fact [1,140]：副 実際に

☐ occur [42,905]：動 起こる，生じる

☐ simultaneously [3,517]：副 同時に（類 concomitantly [678]，concurrently [664]） simultaneously with ～［271］：～と同時に

□ **mycoplasmal**［34］：形 マイコプラズマの

訳 これらの感染は肺炎マイコプラズマ感染と臨床的に区別できないし，実際にマイコプラズマ感染と同時に起こるかもしれない

4 文頭の副詞の文例

文例No			用例数
147	Finally,	最後に，ついに	8,678
148	Interestingly,	興味深いことに，	5,429
149	Additionally,	そのうえ，	3,649
213	Together,	まとめると，	3,624
150	Surprisingly,	驚いたことに，	3,567
151	Recently,	最近，	2,935
152	Conversely,	逆に，	2,468
198	Collectively,	まとめると，	1,826
153	Previously,	以前に，	1,743
154	Consequently,	したがって，	1,197
155	Subsequently,	引き続いて，	662

147 **Finally,** significant progress was achieved in the development of improved vectors for future gene therapy of human diseases.

- □ **finally** [9,234]：副 ついに（類 at last [70]），最後に（類 lastly [399]，terminally [1,098]，ultimately [1,947]，eventually [1,091]）

  Finally, [8,678]：ついに，最後に
- □ **significant** [43,571]：形 著しい（類 marked [6,470]，striking [2,590]，prominent [2,934]），重要な，有意な
- □ **progress** [4,053]：名 進歩，進行（類 progression [12,698]，advance [8,632]）/ 動 進行する
- □ **achieve** [10,535]：動 達成する（類 accomplish [1,840]，attain [806]）
- □ **development** [48,509]：名 開発，発生，発症
- □ **improve** [16,767]：動 改善する
- □ **vector** [12,677]：名 ベクター，媒介物
- □ **future** [4,693]：名 将来
- □ **gene therapy** [2,155]：遺伝子療法

☐ **human** [101,441]：形 ヒトの

☐ **disease** [71,437]：名 疾患

訳 ついに，ヒト疾患の将来の遺伝子治療のために改善されたベクターの開発において著しい進歩が達成された

148 **Interestingly,** infected epithelial cells expressed cytokines that augment γ interferon production.

☐ **interestingly** [5,727]：副 興味深いことに（類 intriguingly [285]）

Interestingly, [5,429]：興味深いことに，

☐ **infected** [20,742]：感染した

☐ **epithelial cell** [10,220]：上皮細胞

☐ **express** [78,977]：動 発現する

☐ **cytokine** [19,780]：名 サイトカイン

☐ **augment** [3,582]：動 増大させる（類 increase [146,670]，elevate [13,039]，up-regulate [3,805]），増強する

☐ **interferon** [6,201]：名 インターフェロン

☐ **production** [27,611]：名 産生

訳 興味深いことに，感染した上皮細胞はγインターフェロン産生を増大させるサイトカインを発現した

149 **Additionally,** six residues exerting significant effects on OCT binding were also found within the putative cleft region.

☐ **additionally** [4,298]：副 そのうえ（類 in addition [27,497]，further [29,238]，furthermore [20,702]，moreover [11,306]）

Additionally, [3,649]：そのうえ，

☐ **residue** [45,837]：名 残基

☐ **exert** [3,821]：動 発揮する

☐ **significant** [43,571]：形 著しい，重要な，有意な

☐ **effect on ～** [23,839]：～に対する影響

☐ **binding** [133,851]：名 結合

☐ **find** [78,775]：動（過 / 過分 found）見つける

☐ **putative** [10,254]：形 推定上の（類 presumptive [718]）

□ **cleft** [1,498]：名 間隙，中裂

□ **region** [77,257]：名 領域（類 area [19,443]，site [99,321]，locus [18,724]）

訳 そのうえ，OCT結合に対する著しい影響を発揮する6つの残基は，また，推定上の間隙領域内に見つけられた

150 **Surprisingly,** the inhibitor-treated cells remained pluripotent despite the absence of leukemia inhibitory factor.

□ **surprisingly** [4,884]：副 驚いたことに

Surprisingly, [3,567]：驚いたことに，

□ **inhibitor** [43,673]：名 阻害剤

□ **～-treated** [10,000]：～に処理された

□ **remain** [26,736]：動 ～のままである

□ **pluripotent** [335]：形 多能性の

□ **absence** [21,500]：名 非存在，存在しないこと

□ **leukemia inhibitory factor** [156]：白血病抑制因子

訳 驚いたことに，阻害剤に処理された細胞は白血病抑制因子の存在がないにもかかわらず多能性のままであった

151 **Recently,** a new hypothesis has been put forward suggesting that metabolic switching is an intrinsic property of skeletal muscle.

□ **Recently,** [2,935]：最近，

□ **hypothesis** [15,247]：名 仮説

□ **put forward** [52]：動 提案する（類 propose [20,814]）

□ **suggest that ～** [96,112]：～ということを示唆する

□ **metabolic** [7,187]：形 代謝性の

□ **switching** [1,779]：名 スイッチング，転換（類 conversion [5,249]，shift [12,913]）

□ **intrinsic** [5,126]：形 内因性の，内在性の（類 endogenous [13,368]）

□ **property** [21,741]：名 性質（類 character [1,245]，nature [6,689]，propensity [1,228]）

□ **skeletal muscle** [4,206]：骨格筋

訳 最近，代謝性スイッチングは骨格筋の内因的な性質であることを示唆する新しい仮説が提案されている

152 **Conversely,** enzymatic and structural analyses showed that the mutation can affect substrate specificity for phosphotyrosine.

- □ **conversely** [2,765]：副 逆に （類 inversely [1,588]，oppositely [134]）
  Conversely, [2,468]：逆に，
- □ **enzymatic** [4,662]：形 酵素的な
- □ **structural** [21,284]：形 構造的な
- □ **analysis** [85,671]：名 （複 analyses） 解析，分析
- □ **show that ～** [74,382]：～ということを示す
- □ **mutation** [70,231]：名 変異
- □ **affect** [32,553]：動 影響する （類 influence [16,903]）
- □ **substrate** [32,684]：名 基質
- □ **specificity** [16,425]：名 特異性
- □ **phosphotyrosine** [1,136]：名 ホスホチロシン

訳 逆に，酵素分析および構造分析はその変異がホスホチロシンに対する基質特異性に影響を与えうることを示した

153 **Previously,** we demonstrated that insulin facilitates muscle protein synthesis in obese Zucker rats.

- □ **Previously,** [1,743]：以前に，
- □ **demonstrate that ～** [45,052]：～ということを実証する
- □ **insulin** [20,353]：名 インスリン
- □ **facilitate** [10,562]：動 促進する （類 accelerate [5,044]，promote [18,750]）
- □ **muscle** [25,413]：名 筋肉
- □ **synthesis** [25,370]：名 合成 （類 production [27,611]）
- □ **obese** [1,906]：形 肥満の （類 fat [6,300]）
- □ **Zucker rat** [108]：ズッカーラット

訳 以前に，我々はインスリンが肥満のズッカーラットにおける筋肉タンパク質合成を促進することを実証した

154 **Consequently,** it is suggested that fetal catecholamines play an important role not only in altering fetal metabolism but also in regulating fetal growth.

□ **consequently** [2,153]：副 したがって，その結果として（類 accordingly [1,249]，therefore [17,502]，hence [2,895]，thus [38,689]）

Consequently, [1,197]：したがって，

□ **it is suggested that ～** [417]：～ということが示唆される

□ **fetal** [9,274]：形 胎児性の

□ **catecholamine** [983]：名 カテコールアミン

□ **play an important role in ～** [5,557]：～の際に重要な役割を果たす

□ **not only ～ but also …** [5,000]：～だけでなく…も

□ **alter** [21,411]：動 変化させる（類 change [63,291]，shift [12,913]，convert [5,751]）

□ **metabolism** [8,783]：名 代謝

□ **regulate** [57,131]：動 調節する

□ **growth** [59,775]：名 成長，増殖

訳 したがって，胎児性のカテコールアミンは胎児の代謝を変える際だけでなく胎児の成長を調節する際にも重要な役割を果たすことが示唆される

155 **Subsequently,** retinoic acid was detected at relatively high levels in the central nervous system of adult rats.

□ **subsequently** [4,526]：副 引き続いて（類 continuously [1,221]）

Subsequently, [662]：引き続いて，

□ **retinoic acid** [2,344]：レチノイン酸

□ **detect** [30,240]：動 検出する

□ **relatively** [9,096]：副 比較的

□ **high** [68,406]：形 高い（反 low [39,773]）

□ **central nervous system** [3,405]：中枢神経系

訳 引き続いて，成体ラットの中枢神経系においてレチノイン酸は比較的高いレベルで検出された

Ⅲ. 形容詞の使い方

形容詞は，主に文の補語や名詞の修飾語として用いられる．形容詞の用法として最も多いのは**形容詞＋名詞**の組合わせであるが，この使い方は比較的容易である．一方，形容詞が補語として用いられたり，名詞を後ろから修飾するために用いられたりする場合には，しばしば形容詞のあとに前置詞が続くという特徴がある．この**形容詞＋前置詞**のパターンは日本語にはないものなので，その使い方には注意を要するだろう．そこでここでは，**形容詞＋前置詞**のパターンを取り上げる．使われる前置詞には，個々の形容詞ごとに決まった組合わせがあるのでそれを習得しておくとよい．

形容詞＋前置詞の文例

文例No			用例数
156	responsible for ～	～に対して責任のある，～の原因である	12,456
157	essential for ～	～のために必須である	11,689
158	important for ～	～のために重要な	7,630
159	necessary for ～	～のために必要な	6,750
185	critical for ～	～にとって決定的に重要な	5,581
160	specific for ～	～に対して特異的な	3,872
161	sufficient for ～	～のために十分な	3,259
162	useful for ～	～のために有用な	2,249
163	crucial for ～	～にとって決定的に重要な	1,536
164	available for ～	～のために利用できる	1,584
165	positive for ～	～に対して陽性の	1,464
166	homozygous for ～	～に対してホモ接合性の	1,186
167	distinct from ～	～と明らかに異なる	3,408
168	different from ～	～と異なる	2,877

文例No			用例数
169	indistinguishable from ～	～と区別できない	1,171
170	independent of ～	～と関係ない，～に依存しない	7,681
054	capable of ～	～できる	7,651
171	dependent on ～	～に依存する	9,107
401	due to ～	～のせいで，～に帰すべき	18,077
036	similar to ～	～に類似している	17,306
172	sensitive to ～	～に感受性の	5,606
173	resistant to ～	～に抵抗性の	4,493
174	identical to ～	～と同一の	3,244
175	comparable to ～	～に匹敵する	2,584
055	adjacent to ～	～に隣接する	2,547
038	susceptible to ～	～に感受性の	2,430
176	attributable to ～	～に起因しうる	1,891
–	specific to ～	～に特異的な	1,622
177	consistent with ～	～と一致する	22,025
178	compatible with ～	～に適合する，～と矛盾しない	987
179	coincident with ～	～と一致する，～と同時に起こる	833
–	comparable with ～	～に匹敵する	753
180	concomitant with ～	～に随伴して，～と同時に	697

156 Lead is thought to be **responsible for** an increased incidence of spontaneous abortion because it readily crosses the placenta.

- □ **lead** [9,294]：名 鉛／動 つながる
- □ **thought to ～** [6,707]：～すると考えられる
- □ **responsible** [12,998]：形 責任のある（類 causative [947]）
  responsible for ～ [12,456]：～に対して責任のある（類 due to [18,077]），～の原因である

□ **increased** [73,539]：増大した

□ **incidence** [8,627]：名 発生率，頻度（類 occurrence [2,697]，frequency [18,970]，rate [59,576]）

□ **spontaneous** [5,990]：形 自発性の

spontaneous abortion [103]：自然流産

□ **readily** [4,144]：副 容易に（類 easily [1,709]），すぐに

□ **cross** [17,837]：動 通過する（類 pass [1,341]），横断する / 名 クロス

□ **placenta** [996]：名 胎盤

訳 鉛は容易に胎盤を通過するので，自然流産の増大した発生率の原因であると考えられている

157 Trace elements such as iron, zinc, copper, chromium, selenium, iodine, fluorine, and manganese have been identified as **essential for** health in humans.

□ **trace element** [68]：微量元素

□ **iron** [10,455]：名 鉄

□ **zinc** [7,513]：名 亜鉛

□ **copper** [3,455]：名 銅

□ **chromium** [269]：名 クロム

□ **selenium** [591]：名 セレン

□ **iodine** [392]：名 ヨウ素

□ **fluorine** [284]：名 フッ素

□ **manganese** [1,027]：名 マンガン

□ **identified as ～** [4,389]：～として同定される

□ **essential** [23,051]：形 必須な（類 necessary [12,049]，mandatory [194]，obligatory [508]，important [45,928]）

essential for ～ [11,689]：～のために必須である

□ **health** [10,469]：名 健康

□ **in humans** [5,894]：ヒトにおいて

訳 鉄，亜鉛，銅，クロム，セレン，ヨウ素，フッ素，マンガンのような微量元素は，ヒトにおいて健康のために必須であると同定されている

158 The residue was found to be **important for** nuclear localization and DNA binding.

- □ **residue** [45,837]：名 残基
- □ **found to ～** [15,859]：～することが見つけられる
- □ **important for ～** [7,630]：～のために重要な
- □ **nuclear localization** [2,049]：核局在化，核移行
- □ **DNA binding** [11,354]：DNA 結合

訳 その残基は，核局在化およびDNA結合のために重要であることが見つけられた

159 The aim of this study was to determine whether stem cells are **necessary for** epidermal renewal.

- □ **aim** [4,574]：名 目的 / 動 目的とする
- □ **determine whether ～** [7,898]：～かどうかを決定する
- □ **stem cell** [7,364]：幹細胞
- □ **necessary** [12,049]：形 必要な（類 essential [23,051]，important [45,928]，fundamental [3,010]）
  necessary for ～ [6,750]：～のために必要な
- □ **epidermal** [5,191]：形 上皮の
- □ **renewal** [624]：名 再生（類 regeneration [2,679]，reproduction [730]）

訳 この研究の目的は，幹細胞が上皮の再生のために必要であるかどうかを決定することであった

160 Monoclonal antibodies **specific for** vaccinia virus have been commercially available.

- □ **monoclonal antibody** [5,131]：単クローン抗体，モノクローナル抗体
- □ **specific for ～** [3,872]：～に対して特異的な
- □ **vaccinia virus** [1,026]：ワクシニアウイルス
- □ **commercially available** [622]：市販されている

訳 ワクシニアウイルスに対して特異的な単クローン抗体が市販されてきた

161 The degradation of local basement membrane is not **sufficient for** distant metastasis of malignant cells.

- □ degradation [11,582]：名 分解
- □ local [9,593]：形 局所の（類 topical [810], focal [4,053] / 反 systemic [7,731]）
- □ basement membrane [1,132]：基底膜
- □ sufficient for ～ [3,259]：～のために十分な
- □ distant metastasis [88]：遠隔転移
- □ malignant [4,511]：形 悪性の

訳 局所の基底膜の分解は，悪性細胞の遠隔転移にとって十分ではない

162 This model may be **useful for** studying the role of posttranslational modification.

- □ model [72,382]：名 モデル / 動 モデル化する
- □ useful for ～ [2,249]：～のために有用な
- □ study the role of ～ [872]：～の役割を研究する
- □ posttranslational modification [567]：翻訳後修飾

訳 このモデルは，翻訳後修飾の役割を研究するために有用であるかもしれない

163 Angiogenesis is **crucial for** tumor growth.

- □ angiogenesis [4,263]：名 血管新生
- □ crucial [4,757]：形 決定的に重要な（類 critical [21,712], definitive [1,252], important [45,928], conclusive [176]）

  crucial for ～ [1,536]：～にとって決定的に重要な
- □ tumor growth [2,043]：腫瘍増殖，腫瘍成長

訳 血管新生は，腫瘍増殖にとって決定的に重要である

164 An array of diagnostic tests is now **available for** evaluation of patients with evident or suspected cardiovascular disease.

- □ an array of ～ [499]：一連の～

☐ **diagnostic** [4,965]：形 診断の

diagnostic test [334]：診断検査

☐ **now** [10,465]：副 今（類 presently [557], currently [4,311]）

☐ **available for 〜** [1,584]：〜のために利用できる

☐ **evaluation** [5,663]：名 評価（類 assessment [5,256], estimation [961]）

☐ **patient with 〜** [43,273]：〜の患者

☐ **evident** [2,899]：形 明らかな

☐ **suspect** [1,509]：動 疑う，思う

☐ **cardiovascular disease** [1,853]：循環器疾患，心血管疾患

訳 一連の診断検査は，今，明らかなあるいは疑いのある循環器疾患の患者の評価のために利用できる

165 Seven of 95 specimens tested **positive for** HIV antibodies.

☐ **specimen** [5,996]：名 検体

☐ **test** [45,529]：動 テストする，検査する（類 examine [39,969], study [167,914], investigate [30,619]），検査で〜と出る / 名 テスト，検査

☐ **positive** [26,646]：形 陽性の，正の（類 active [33,441]）/ 名 陽性であること

positive for 〜 [1,464]：〜に対して陽性の

☐ **HIV** [29,027]：名 ヒト免疫不全ウイルス

☐ **antibody** [36,724]：名 抗体

訳 95 検体のうち 7 つが HIV 抗体に対して検査で陽性と出た

166 Severe developmental abnormalities have been reported in mice **homozygous for** this enzyme deficiency.

☐ **severe** [12,151]：形 重篤な（類 serious [1,674]），重症の，激しい

☐ **developmental** [7,562]：形 発生の，発生上の

☐ **abnormality** [7,645]：名 異常

☐ **report** [54,972]：動 報告する / 名 報告

☐ **homozygous** [4,049]：形 ホモ接合性の

homozygous for 〜 [1,186]：〜に対してホモ接合性の

☐ **enzyme** [49,164]：名 酵素

□ **deficiency** [6,979]：名 欠損，欠損症（類 deficit [3,988]，defect [18,182]），欠乏

訳 重篤な発生異常が，この酵素欠損症に対してホモ接合性であるマウスにおいて報告されている

167 This secretion signal seems to be **distinct from** the N-terminal signal peptide.

□ **secretion** [12,758]：名 分泌，分泌物

□ **signal** [40,955]：名 シグナル，信号 / 動 信号を送る

□ **seem to ～** [3,153]：～するように思われる

□ **distinct from ～** [3,408]：～と明らかに異なる

□ **N-terminal** [11,936]：形 N 末端の

□ **signal peptide** [871]：シグナルペプチド

訳 この分泌シグナルは，N 末端シグナルペプチドとは明らかに異なっているように思われる

168 Survival was significantly **different from** what was expected for an age- and sex-matched population.

□ **survival** [28,139]：名 生存，生存率

□ **significantly** [48,939]：副 有意に，著しく

□ **different from ～** [2,877]：～と異なる

□ **expected for ～** [520]：～に対して予想される，～に対して期待される

□ **sex-matched** [192]：性別の一致した

□ **population** [28,214]：名 集団，人口

訳 生存率は，年齢と性別の一致した集団に対して予想されたものと有意に異なっていた

169 Mice lacking one of these receptors are **indistinguishable from** their wild-type littermates.

□ **lack** [23,888]：動 欠く（類 lose [4,468]）/ 名 欠如

□ **receptor** [125,789]：名 受容体

□ **indistinguishable from ～** [1,171]：～と区別できない

☐ **wild-type** [39,564]：野生型の

☐ **littermate** [1,364]：名 同腹仔

訳 これらの受容体のひとつを欠いているマウスは，野生型の同腹仔と区別できない

170 Major congenital anomalies were confirmed in 3 to 4% of all newborn infants, **independent of** ethnic group.

☐ **major** [30,348]：形 主要な

☐ **congenital** [1,980]：形 先天性の

☐ **anomaly** [914]：名 異常

☐ **confirm** [17,664]：動 確認する（類 verify [1,730], validate [2,784], ascertain [1,152]）

☐ **all** [70,258]：形 すべての（類 whole [8,514], entire [4,681], overall [14,048]）

☐ **newborn infant** [38]：新生児

☐ **independent of 〜** [7,681]：〜と関係ない，〜に依存しない

☐ **ethnic group** [415]：民族

訳 主要な先天性異常が，民族に関係なく，すべての新生児の３から４%において確認された

171 The therapeutic effects on proteinuria, renal pathologic features, and survival rate were found to be **dependent on** the drug dose.

☐ **therapeutic** [10,491]：形 治療の

therapeutic effect [339]：治療効果

☐ **proteinuria** [526]：名 蛋白尿

☐ **renal** [11,739]：形 腎臓の

☐ **pathologic** [1,842]：形 病理学的な

☐ **feature** [15,065]：名 特徴（類 characteristic [14,974]）

☐ **survival rate** [1,766]：生存率

☐ **found to 〜** [15,859]：〜することが見つけられる

☐ **dependent on 〜** [9,107]：〜に依存する

☐ **drug dose** [49]：薬剤投与量

訳 蛋白尿，腎臓の病理学的特徴，および生存率に対する治療効果は，薬剤投与量に依存することが見つけられた

172 Diagnostic tests and other clinical evaluations help to identify tumors that are **sensitive to** this particular drug.

- □ **diagnostic test** [334]：診断検査
- □ **other** [79,407]：形 他の / 名 他のもの
- □ **clinical** [31,256]：形 臨床の
- □ **evaluation** [5,663]：名 評価
- □ **help** [6,063]：動 役立つ，助ける（類 support [28,621]，assist [3,032]）

  help to ～ [1,718]：～するのに役立つ（to 不定詞であるが，help の後は例外的に to が省略されることの方が多い）
- □ **identify** [73,456]：動 同定する
- □ **tumor** [61,098]：名 腫瘍
- □ **sensitive to ～** [5,606]：～に感受性の
- □ **particular** [7,937]：形 特定の
- □ **drug** [23,365]：名 薬剤（類 agent [18,880]，medication [2,850]）

訳 診断検査および他の臨床的評価は，この特定の薬剤に感受性である腫瘍を同定するのに役立つ

173 The two groups of strains were extremely **resistant to** the bactericidal activity of serum.

- □ **group** [69,369]：名 グループ，群（類 population [28,214]），基 / 動 グループ化する
- □ **strain** [33,247]：名 株，菌株
- □ **extremely** [2,414]：副 極度に（類 unusually [935]，markedly [7,503]）
- □ **resistant to ～** [4,493]：～に抵抗性の
- □ **bactericidal activity** [245]：殺菌性活性
- □ **serum** [21,260]：名 血清

訳 2つのグループの菌株は，血清の殺菌性活性に極度に抵抗性であった

174 The pattern of expression of proapoptotic genes was nearly **identical to** the pattern of apoptotic genes induced by tumor necrosis factor α.

- □ **pattern** [33,282]：名 パターン，様式
- □ **expression** [154,475]：発現
- □ **proapoptotic** [1,133]：形 アポトーシス促進性の
- □ **nearly** [5,643]：副 ほとんど（類 almost [4,947]，virtually [2,089]，mostly [1,734]）
- □ **identical** [9,552]：形 同一の，等しい（類 same [24,047]，equal [2,730]，equivalent [5,237]，comparable [6,413]）
  - identical to ～ [3,244]：～と同一の
- □ **apoptotic** [8,348]：形 アポトーシスの，アポトーシス性の
- □ **induced by ～** [13,321]：～によって誘導される
- □ **tumor necrosis factor** [5,278]：腫瘍壊死因子（TNF）

訳 アポトーシス促進性遺伝子の発現のパターンは，腫瘍壊死因子αによって誘導されるアポトーシス遺伝子のパターンとほとんど同一であった

175 The results obtained by this rapid, reproducible, and noninvasive method, computed tomography, were **comparable to** those obtained by histology.

- □ **obtain** [21,689]：動 得る（類 acquire [5,529]，gain [6,383]）
- □ **rapid** [15,767]：形 急速な，迅速な（類 prompt [401]，quick [177]）
- □ **reproducible** [1,187]：形 再現性のある
- □ **noninvasive** [1,384]：形 非侵襲性の
- □ **method** [32,975]：名 方法（類 procedure [8,622]，manner [13,342]）
- □ **computed tomography** [865]：コンピュータ断層撮影（CT）
- □ **comparable** [6,249]：形 匹敵する，同等の（類 equivalent [5,237]，equal [2,730]，same [24,047]）
  - comparable to ～ [2,580]：～に匹敵する（類 due to [18,077]）
- □ **histology** [1,281]：名 組織学，組織診断

訳 この迅速で再現性があり非侵襲性の方法，コンピュータ断層撮影，によって得られた結果は，組織診断によって得られたそれらに匹敵した

176 Randomized double-blind trials should be carried out to detect common adverse reactions **attributable to** this vaccine.

- □ **randomized** [5,255]：形 無作為化される，ランダム化される
- □ **double-blind** [1,351]：形 二重盲検の
- □ **trial** [13,519]：名 治験，試行（類 attempt [3,991]，study [167,914]，test [45,529]）
- □ **carry out** [3,749]：行う，実行する
- □ **detect** [30,240]：動 検出する
- □ **common** [19,789]：形 よくある，共通の（類 general [10,998]，usual [1,028]）
- □ **adverse reaction** [125]：有害反応
- □ **attributable** [2,207]：形 起因しうる（類 causative [947]）
  attributable to ～ [1,891]：～に起因しうる（類 due to [18,077]）
- □ **vaccine** [10,085]：名 ワクチン

訳 無作為化された二重盲検治験が，このワクチンに起因しうるよくある有害反応を検出するために実行されるべきである

177 These findings are **consistent with** the hypothesis that the functional maturation of the primate visual brain proceeds in a hierarchical manner.

- □ **finding** [35,256]：名 知見
- □ **consistent with ～** [22,025]：～と一致する（類 coincident with [833]，corresponding to [3,708]）
- □ **the hypothesis that ～** [7,746]：～という仮説
- □ **functional** [34,575]：形 機能的な
- □ **maturation** [5,412]：名 成熟
- □ **primate** [2,966]：名 霊長類
- □ **visual** [8,105]：形 視覚の
- □ **brain** [28,011]：名 脳
- □ **proceed** [2,472]：動 進行する（類 progress [4,053]，advance [8,632]）
- □ **hierarchical** [721]：形 階層的な
- □ **manner** [13,342]：名 様式（類 fashion [13,342]，mode [6,316]，way [5,601]）

訳 これらの知見は，霊長類の視覚脳の機能的成熟が階層的な様式で進行するという仮説と一致する

178 These in vitro findings were **compatible with** in vivo observations of synapse growth and elimination in living tadpoles of Xenopus laevis.

- □ in vitro [41,334]：形 試験管内の / 副 試験管内で
- □ finding [35,256]：名 知見
- □ compatible [1,340]：形 矛盾しない，適合する（類 suitable [1,857]）
  compatible with ～ [987]：～と矛盾しない，～に適合する（類 consistent with [22,025]，coincident with [833]）
- □ in vivo [41,680]：形 生体内の，副 生体内で
- □ observation [14,035]：名 観察
- □ synapse [6,527]：名 シナプス
- □ growth [59,775]：名 成長，増殖（類 proliferation [19,120]，replication [21,221]）
- □ elimination [2,643]：名 消失（類 disappearance [644]），除去（類 removal [5,555]，abstraction [466]）
- □ living [3,761]：形 生きている（類 alive [543]）
- □ tadpole [232]：名 オタマジャクシ
- □ Xenopus laevis [792]：アフリカツメガエル

訳 これらの試験管内の知見は，生きているアフリカツメガエルのオタマジャクシにおけるシナプスの成長と消失の生体内の観察と矛盾しなかった

179 The activation of mitogen-activated protein kinase was **coincident with** the age-dependent increase in amyloid deposition and loss of synaptophysin.

- □ activation [82,818]：名 活性化（類 potentiation [2,148]，transactivation [3,787]）
- □ mitogen-activated protein kinase [3,682]：マイトジェン活性化プロテインキナーゼ
- □ coincident [1,079]：形 一致する，同時に起こる
  coincident with ～ [833]：～と一致する（類 consistent with [22,025]，corresponding to [3,708]），～と同時に起こる

III 形容詞の使い方

☐ **age-dependent** [462]：年齢依存的な

☐ **increase in ～** [35,718]：～の増大

☐ **amyloid deposition** [147]：アミロイド沈着

☐ **loss** [28,751]：名 喪失, 減少（類 deficit [3,988], defect [18,182], lack [23,888], decrease [51,139]）

☐ **synaptophysin** [289]：名 シナプトフィジン

訳 マイトジェン活性化プロテインキナーゼの活性化は，アミロイド沈着の年齢依存的な増大およびシナプトフィジンの喪失と同時に起こった

180 **Concomitant with** the reduction in spheroid number, the mutant mice showed increased viability and an improved behavioral phenotype.

☐ **concomitant** [3,017]：形 随伴性の，同時の（類 concurrent [1,569], coincident [1,079], simultaneous [2,933]）

concomitant with ～ [697]：～に随伴して，～と同時に

☐ **reduction** [23,591]：名 低下, 減少（類 decrease [51,139], decline [6,411], attenuation [2,259], fall [2,820]）

☐ **spheroid** [204]：名 スフェロイド，球状体 / 形 球状の

☐ **number** [39,607]：名 数 / 動 番号を付ける

☐ **mutant** [79,727]：形 変異の / 名 変異体

☐ **show** [148,875]：動 示す（類 present [53,747], indicate [73,855], exhibit [27,339]）

☐ **increased** [73,539]：増大した

☐ **viability** [3,419]：名 生存率，生存能（類 survival [28,139]）

☐ **improve** [16,767]：動 改善する

☐ **behavioral** [4,542]：形 行動上の

☐ **phenotype** [23,661]：名 表現型

訳 スフェロイドの数の低下に随伴して，変異マウスは増大した生存率および改善された行動上の表現型を示した

Ⅳ. 名詞の使い方

名詞は文の最も主要な構成要素であり，主語，目的語，補語や句などの中心となるものとして用いられる．名詞にはさまざまな修飾語がつくが，特に形容詞句による後ろからの修飾は，日本語にはない文の組立なので使い方に気を付けよう．このような形容詞句はしばしば前置詞に導かれ，そのとき使われる前置詞は**of**が圧倒的に多い．しかし，名詞によってはそれ以外の前置詞もしばしば用いられる．また，同格の**that節**も一部の名詞だけに用いられる特別なものである．ここでは，そのようなof以外の**名詞＋前置詞**および**名詞＋同格のthat節**の特徴的な組合せに着目して示す．

1 名詞＋前置詞の文例

文例No			用例数
181	information about ～	～に関する情報	1,714
182	… hours after ～	～の後…時間	1,606
183	protection against ～	～に対する保護	1,834
－	antibody against ～	～に対する抗体	1,934
184	activity against ～	～に対する活性	1,311
185	defense against ～	～に対する防御	713
186	age at ～	～時の年齢	1,458
284	interaction between ～	～の間の相互作用	11,513
187	relationship between ～	～の間の関連性	6,532
188	association between ～	～の間の関連	4,561
189	difference between ～	～の間の違い	5,511
190	correlation between ～	～の間の相関	3,834
191	link between ～	～の間の関連	2,406
－	relation between ～	～の間の関連	1,391
192	mechanism by which ～	（それによって）～である機構	6,380

文例No			用例数
193	activation by ～	～による活性化	3,749
194	expression by ～	～による発現	2,645
−	activity by ～	～による活性	2,418
195	inhibition by ～	～による抑制	2,097
196	evidence for ～	～の証拠	7,029
005	model for ～	～のためのモデル	6,728
197	method for ～	～のための方法	4,979
198	basis for ～	～に対する基礎，～の基礎	4,886
199	requirement for ～	～に対する要求性	4,857
200	affinity for ～	～に対する親和性	4,315
059	risk for ～	～のリスク	4,282
201	target for ～	～のための標的	4,609
−	site for ～	～のための部位	4,768
202	need for ～	～の必要性	4,087
203	implication for ～	～のための意味	3,454
073	receptor for ～	～に対する受容体	3,144
204	gene for ～	～の遺伝子	3,131
205	strategy for ～	～のための戦略	2,739
206	therapy for ～	～の治療	2,467
207	tool for ～	～のための道具	2,465
208	system for ～	～のためのシステム	2,458
286	substrate for ～	～に対する基質	2,770
−	value for ～	～に対する価値	2,704
209	potential for ～	～のための潜在能	2,201
210	pathway for ～	～の経路	2,129
211	adjustment for ～	～に対する調整	1,932
212	support for ～	～に対する支持	1,609

文例No			用例数
087	approach for ～	～のためのアプローチ	1,923
213	explanation for ～	～に対する説明	1,582
214	search for ～	～の探索	1,455
215	specificity for ～	～に対する特異性	1,377
216	data from ～	～からのデータ	3,851
217	release from ～	～からの放出	1,916
218	sample from ～	～からのサンプル	1,818
025	increase in ～	～の増大	28,190
190	change in ～	～の変化	22,681
219	difference in ～	～の違い	11,719
025	decrease in ～	～の低下	11,127
220	reduction in ～	～の低下	10,211
194	insight into ～	～への洞察	3,510
221	incorporation into ～	～への取り込み	703
222	data on ～	～に関するデータ	2,130
223	impact on ～	～に対する影響	1,831
224	study on ～	～に関する研究	1,774
225	information on ～	～に関する情報	1,758
226	influence on ～	～に対する影響	1,094
227	dependence on ～	～への依存性	955
228	signaling through ～	～を経るシグナル伝達	1,145
229	resistance to ～	～に対する抵抗性	5,635
–	approach to ～	～するアプローチ	5,070
–	antibody to ～	～に対する抗体	4,123
195	sensitivity to ～	～に対する感受性	3,966
230	susceptibility to ～	～に対する感受性	3,082
231	similarity to ～	～に対する類似性	2,596

文例No			用例数
232	homology to ～	～に対する相同性	2,496
079	ability to *do*	～する能力	15,068
–	method to *do*	～する方法	2,931
034	patient with ～	～の患者	43,273
–	interaction with ～	～との相互作用	10,878
233	treatment with ～	～による処理，～による治療	8,802
234	complex with ～	～との複合体	5,345
–	association with ～	～との結合	5,329
–	infection with ～	～による感染	3,348

181 The present study provides biochemical **information about** the substrate specificity of the enzyme.

- □ **present** [41,922]：形 現在の，存在する / 動 提示する
- □ **provide** [56,723]：動 提供する（類 offer [4,619]，supply [2,088]），与える
- □ **biochemical** [9,194]：形 生化学的な
- □ **information** [15,768]：名 情報
  - **information about ～** [1,714]：～に関する情報
- □ **substrate** [32,684]：名 基質
- □ **specificity** [16,425]：名 特異性
- □ **enzyme** [49,164]：名 酵素

訳 現在の研究は，その酵素の基質特異性に関する生化学的情報を提供する

182 Per1 mRNA levels peaked at 2 **hours after** stimulation.

- □ **peak** [9,497]：動 ピークになる，ピークに達する / 名 ピーク
- □ …**hours after ～** [1,606]：～の後…時間
- □ **stimulation** [21,013]：名 刺激

訳 Per1 メッセンジャーRNA レベルは，刺激の後 2 時間でピークになった

183 **Protection against** apoptotic insults may arise through the inhibition of a proapoptotic protein.

☐ **protection** [8,224]：名 保護

protection against ～ [1,834]：～に対する保護

☐ **apoptotic** [8,348]：形 アポトーシス性の，アポトーシスの

☐ **insult** [959]：名 侵襲，傷害（類 injury [15,721]，aggression [437]）

☐ **arise** [6,458]：動 起こる，生じる（類 occur [42,905]，take place [1,287]，emerge [4,376]）

☐ **inhibition** [37,129]：名 抑制

☐ **proapoptotic** [1,133]：形 アポトーシス促進性の

訳 アポトーシス性侵襲に対する保護は，アポトーシス促進性タンパク質の抑制を経て起こるかもしれない

184 Interleukin-4 cytotoxin exhibited remarkable antitumor **activity against** ovarian tumors in immunodeficient animals.

☐ **interleukin** [8,723]：名 インターロイキン

☐ **cytotoxin** [246]：名 サイトトキシン

☐ **exhibit** [27,339]：動 示す

☐ **remarkable** [1,492]：形 顕著な，注目すべき

☐ **antitumor** [2,537]：形 抗腫瘍性の

☐ **activity against ～** [1,311]：～に対する活性

☐ **ovarian** [3,763]：形 卵巣の

☐ **tumor** [61,098]：名 腫瘍

☐ **immunodeficient** [680]：形 免疫不全の

訳 インターロイキン4サイトトキシンは，免疫不全動物における卵巣腫瘍に対する顕著な抗腫瘍活性を示した

185 **Defense against** host-derived reactive oxygen species has previously been reported as critical for intracellular replication.

☐ **defense against ～** [713]：～に対する防御

- □ **host-derived** [133]：形 宿主由来の
- □ **reactive oxygen species** [1,651]：活性酸素種
- □ **previously** [34,828]：副 以前に
- □ **report** [54,972]：動 報告する / 名 報告
- □ **critical for ～** [5,581]：～にとって決定的に重要である
- □ **intracellular** [20,566]：形 細胞内の
- □ **replication** [21,221]：名 複製（類 proliferation [19,120], growth [59,775]）

訳 宿主由来の活性酸素種に対する防御は，細胞内複製にとって決定的に重要であると以前に報告されている

186 The mean **age at** onset of symptoms was 6.7 years and at diagnosis was 7.3 years.

- □ **mean** [24,324]：名 平均
- □ **age** [36,270]：名 年齢 / 動 加齢する
  - **age at ～** [1,458]：～時の年齢
- □ **onset** [9,566]：名 開始，発症（類 initiation [10,562], start [6,641]）
- □ **symptom** [9,302]：名 症状（類 manifestation [1,760], sign [2,848], indication [1,386]）
- □ **diagnosis** [8,763]：名 診断

訳 症状の発生時の平均年齢は6.7歳，そして診断時の平均年齢は7.3歳であった

187 A complex **relationship between** food and health in our society is intrinsically linked to the obesity epidemic.

- □ **complex** [80,256]：形 複雑な（類 complicated [1,123]），名 複合体，動 複合体を形成する
- □ **relationship** [13,962]：名 関連性，関係（類 relation [4,419], association [24,687], relevance [1,908], correlation [9,923], link [28,890]）
  - **relationship between ～** [6,532]：～の間の関連性
- □ **food** [4,947]：名 食物
- □ **health** [10,469]：名 健康
- □ **society** [792]：名 社会

□ **intrinsically** [554]：副 内因的に（類 endogenously [660]），本質的に（類 essentially [2,269]）

□ **link** [28,890]：動 関連づける，連結する（類 connect [2,458]，couple [17,384]，relate [38,355]，correlate [21,378]，associate [82,102]，bind [187,860]），名 関連

linked to ～ [5,539]：～と関連する（類 associated with [54,439]，related to [10,949]，correlated with [9,967]），～に連結される

□ **obesity** [3,240]：名 肥満（類 adiposity [455]，fatness [107]）

□ **epidemic** [1,138]：名 流行病（類 adiposity [455]，fatness [107] / 形 流行の

訳 我々の社会における食物と健康の間の複雑な関連性は，肥満の流行に内因的に関連する

188 To date, no studies have evaluated **associations between** human leukocyte antigen class II markers and posttreatment chronic Lyme disease.

□ **to date** [2,742]：今までに（類 until now [246]）

□ **evaluate** [19,786]：動 評価する

□ **association between ～** [4,561]：～の間の関連

□ **human leukocyte antigen** [302]：ヒト白血球抗原（HLA）

□ **antigen** [23,859]：名 抗原

□ **class** [25,116]：名 クラス，分類

□ **marker** [18,107]：名 マーカー

□ **posttreatment** [285]：形 治療後の

□ **chronic** [14,659]：形 慢性の

□ **Lyme disease** [711]：ライム病

訳 今までに，ヒト白血球抗原クラスIIマーカーと治療後の慢性ライム病の間の関連を評価した研究はない

189 There were no identifiable **differences between** adults and children with respect to the sensitivity and specificity of serologic tests for celiac disease.

□ **identifiable** [449]：形 同定可能な

- □ difference between ～ [5,511]：～の間の違い
- □ adult [20,020]：名 大人
- □ with respect to ～ [4,188]：～に関して
- □ sensitivity [17,139]：名 感受性（類 susceptibility [7,728]）
- □ specificity [16,425]：名 特異性
- □ serologic [548]：形 血清学的な（類 serological [301]）
- □ celiac disease [167]：セリアック病

訳 セリアック病に対する血清学的試験の感受性と特異性に関して大人と子供の間で同定できる違いはなかった

190 There was a strong **correlation between** changes in the number of tumor cells and clinical status.

- □ correlation [9,923]：名 相関（類 relation [4,419], relationship [13,962], association [24,687], relevance [1,908], link [28,890]）
  correlation between ～ [3,834]：～の間の相関
- □ change in ～ [30,981]：～の変化
- □ tumor cell [7,332]：腫瘍細胞
- □ clinical [31,256]：形 臨床の
- □ status [8,697]：名 状態（類 condition [30,859], situation [1,437]）

訳 腫瘍細胞の数の変化と臨床状態の変化の間に強い相関があった

191 This discovery suggests a possible **link between** aging and stem cell dysfunction.

- □ discovery [3,618]：名 発見（類 finding [35,256], identification [10,858], observation [14,035]）
- □ suggest [126,371]：動 示唆する
- □ possible [14,443]：形 可能な, ありうる（類 feasible [1,123], probable [873]）
- □ link [28,890]：名 関連（類 relation [4,419], relationship [13,962], association [24,687], relevance [1,908], correlation [9,923]）, 動 関連づける，連結する
  link between ～ [2,406]：～の間の関連
- □ aging [2,738]：名 老化，加齢
- □ stem cell [7,364]：幹細胞

□ dysfunction [6,236]：名 機能障害

訳 この発見は，老化と幹細胞機能障害の間のありうる関連を示唆する

192

Investigations of leprosy provide new insights into the **mechanisms by which** the innate immune system contributes to host defense against infection.

□ investigation [6,218]：名 研究，調査

□ leprosy [161]：名 らい病

□ provide [56,723]：動 提供する，与える

□ insight [7,531]：名 洞察

□ mechanism by which ～ [6,380]：（それによって）～である機構

□ innate immune system [337]：自然免疫系

□ contribute to ～ [19,994]：～に寄与する

□ host [19,338]：名 宿主

□ defense against ～ [713]：～に対する防御

□ infection [44,601]：名 感染

訳 らい病の研究は，自然免疫系が感染に対する宿主防御に寄与する機構への新しい洞察を提供する

193

Insulated from the external environment, the pineal gland responds to discontinuous **activation by** the circadian clock.

□ insulate [195]：動 遮断する（類 shield [504]，mask [1,177]）

□ external [3,834]：形 外部の（類 outside [2,792] / 反 internal [6,287]）

□ environment [7,753]：名 環境（類 circumstance [761]）

□ pineal gland [111]：松果体

□ respond to ～ [5,457]：～に応答する，～に反応する

□ discontinuous [338]：形 非連続的な

□ activation by ～ [3,749]：～による活性化

□ circadian clock [587]：概日時計

訳 外部環境から遮断されると，松果体は概日時計による非連続的な活性化に応答する

194 In order to gain insights into acute rejection, we investigated MHC class II **expression by** the allograft.

□ **in order to 〜** [4,148]：〜するために

□ **gain** [6,383]：動 得る（類 obtain [21,689], acquire [5,529], get [280]）/ 名 獲得

□ **insight** [7,531]：名 洞察

□ **acute** [17,897]：形 急性の

□ **rejection** [7,059]：名 拒絶

□ **investigate** [30,619]：動 精査する

□ **MHC** [9,764]：MHC，主要組織適合複合体（major histocompatibility complex）

□ **expression by 〜** [2,645]：〜による発現

□ **allograft** [5,871]：名 同種移植

訳 急性拒絶への洞察を得るために，我々は同種移植によるMHCクラスIIの発現を精査した

195 We sought to evaluate the genetic contribution to sensitivity to cell growth **inhibition by** anticancer agents.

□ **sought to 〜** [2,674]：〜しようと努めた

□ **evaluate** [19,786]：動 評価する

□ **genetic** [28,358]：形 遺伝的な（類 heritable [571]）

□ **contribution** [7,770]：名 寄与

□ **sensitivity to 〜** [3,966]：〜に対する感受性

□ **cell growth** [4,368]：細胞増殖

□ **inhibition by 〜** [2,097]：〜による抑制

□ **anticancer** [1,282]：形 抗癌性の

□ **agent** [18,880]：名 薬剤，作用物質

訳 我々は，抗癌剤による細胞増殖抑制に対する感受性への遺伝的寄与を評価しようと努めた

196 These data collectively offer compelling **evidence for** the transmembrane segment providing a new model of this protein.

- □ **collectively** [2,430]：副 全体として，まとめると
- □ **offer** [4,619]：動 提供する（類 provide [56,723]，supply [2,088]）
- □ **compelling** [532]：形 説得力のある，抗しがたい
- □ **evidence for ～** [7,029]：～の証拠
- □ **transmembrane segment** [636]：膜貫通領域
- □ **provide** [56,723]：動 提供する，与える

訳 これらのデータは，全体としてこのタンパク質の新たなモデルを与える膜貫通領域の説得力のある証拠を提供する

197 We have developed a quantitative **method for** determining protein interactions in cells.

- □ **develop** [38,325]：動 開発する（類 exploit [1,885]），発症する，発達する
- □ **quantitative** [9,052]：形 定量的な（反 qualitative [1,100]）
- □ **method for ～** [4,979]：～のための方法
- □ **determine** [58,574]：動 決定する
- □ **interaction** [63,464]：名 相互作用

訳 我々は，細胞内でのタンパク質相互作用を決定するための定量的な方法を開発した

198 Collectively, these results may provide a cellular and biochemical **basis for** heart failure.

- □ **collectively** [2,430]：副 まとめると（類 taken together [5,918]，in summary [1,701]，in conclusion [2,471]），全体として
- □ **provide** [56,723]：動 提供する，与える
- □ **cellular** [29,968]：形 細胞の，細胞性の
- □ **biochemical** [9,194]：形 生化学的な
- □ **basis** [17,139]：名 基礎
  - **basis for ～** [4,886]：～に対する基礎，～の基礎
- □ **heart failure** [4,001]：心不全

IV 名詞の使い方

訳 まとめると，これらの結果は心不全に対する細胞性および生化学的基礎を提供するかもしれない

199 Dextrose infusion leads to a decrease in the **requirement for** amino acid oxidation as an energy source.

□ dextrose ［137］：名 ブドウ糖

□ infusion ［6,575］：名 注入，点滴（類 injection ［12,342］，transfusion ［1,484］）

□ lead to ～ ［31,651］：～につながる

□ decrease in ～ ［13,282］：～の低下

□ requirement ［7,766］：名 要求性（類 need ［12,039］，demand ［1,297］，request ［483］，necessity ［344］，claim ［654］）

requirement for ～ ［4,857］：～に対する要求性

□ amino acid ［33,123］：アミノ酸

□ oxidation ［7,149］：名 酸化

□ energy ［15,976］：名 エネルギー

□ source ［10,013］：名 源，出所（類 origin ［7,942］）

訳 ブドウ糖注入は，エネルギー源としてアミノ酸酸化に対する要求性の低下につながる

200 Zinc is bound to metallothioneins synthesized in the liver and has a strong **affinity for** red cells and plasma proteins.

□ zinc ［7,513］：名 亜鉛

□ bound to ～ ［6,623］：～に結合した（類 linked to ～ ［5,539］）

□ metallothionein ［361］：名 メタロチオネイン

□ synthesize ［9,021］：動 合成する（類 produce ［38,705］，generate ［28,107］）

□ strong ［11,381］：形 強い，強力な（類 powerful ［2,820］，robust ［3,181］，intensive ［2,341］，potent ［11,330］）

□ affinity ［26,854］：名 親和性

affinity for ～ ［4,315］：～に対する親和性

□ red cell ［646］：赤血球

□ plasma ［22,942］：名 血漿

訳 亜鉛は，肝臓において合成されるメタロチオネインに結合し，そして赤血球および血漿タンパク質に対する強い親和性を持つ

201 This enzyme appears to play a role in controlling cell cycles and can be an important **target for** new chemotherapy agents.

- ☐ **enzyme** [49,164]：名 酵素
- ☐ **appear to ～** [21,779]：～するように思われる
- ☐ **play a role in ～** [6,552]：～の際に役割を果たす
- ☐ **control** [78,085]：動 制御する（類 regulate [57,131], modulate [13,347], mediate [71,309]）/ 名 対照群，コントロール，制御
- ☐ **cell cycle** [13,210]：細胞周期
- ☐ **important** [45,928]：形 重要な
- ☐ **target for ～** [4,609]：～のための標的
- ☐ **chemotherapy** [5,426]：名 化学療法
- ☐ **agent** [18,880]：名 薬剤，作用物質

訳 この酵素は，細胞周期を調節する際に役割を果たすように思われる，そして新しい化学療法剤のための重要な標的でありうる

202 These results highlight the **need for** further studies of the effect of isoflavones on bone.

- ☐ **highlight** [2,970]：動 強調する（類 emphasize [1,584], underscore [1,128]）
- ☐ **need** [12,039]：名 必要性，動 必要とする
  need for ～ [4,087]：～の必要性
- ☐ **further** [29,238]：形 さらに進んだ（far の比較級）/ 副 さらに
- ☐ **effect of … on ～** [6,000]：…の～に対する効果
- ☐ **isoflavone** [413]：名 イソフラボン
- ☐ **bone** [16,753]：名 骨

訳 これらの結果は，イソフラボンの骨に対する効果のさらに進んだ研究の必要性を強調する

203 These findings have **implications for** the design of future clinical trials in patients with prostate cancer.

- ☐ **finding** [35,256]：名 知見

IV 名詞の使い方

- □ **implication** [6,192]：名 潜在的重要性, 意味, 影響（類 importance [8,666], significance [4,937]）
  - **implication for ～** [3,454]：～のための意味
- □ **design** [14,086]：名 設計, 計画（類 plan [1,911]）/ 動 計画する, 設計する
- □ **future** [4,693]：名 将来
- □ **clinical trial** [3,713]：臨床治験
- □ **patient with ～** [43,273]：～の患者
- □ **prostate cancer** [4,808]：前立腺癌

訳 これらの知見は, 前立腺癌の患者の将来の臨床治験の設計のための潜在的重要性を持つ

204 A strong candidate **gene for** Alzheimer's disease was identified in a region at chromosome 12.

- □ **candidate** [7,050]：名 候補
- □ **gene for ～** [3,131]：～の遺伝子
- □ **Alzheimer's disease** [2,946]：アルツハイマー病
- □ **identified in ～** [5,871]：～において同定される
- □ **region** [77,257]：名 領域
- □ **chromosome** [24,029]：名 染色体

訳 アルツハイマー病の強力な候補遺伝子が, 第12染色体のある領域に同定された

205 The authors present a novel **strategy for** generating and analyzing comprehensive genetic-interaction maps.

- □ **author** [5,078]：名 著者
- □ **present** [53,747]：動 提示する, 示す / 形 存在する, 現在の
- □ **novel** [30,313]：形 新規の, 新しい
- □ **strategy for ～** [2,739]：～のための戦略
- □ **generate** [28,107]：動 作成する, 産生する
- □ **analyze** [18,249]：動 分析する, 解析する
- □ **comprehensive** [1,968]：形 包括的な, 網羅的な
- □ **genetic interaction** [556]：遺伝子相互作用

□ **map** [18,094]：名 地図，マップ / 動 位置づける

訳 著者らは，包括的な遺伝子相互作用地図を作成し，そして分析するための新規の戦略を提示する

206 Recent advances in understanding the molecular mechanisms of the differentiation process will hopefully lead to more effective **therapies for** lung cancer.

□ **recent** [18,452]：形 最近の（類 previous [16,852]）

□ **advance** [8,632]：名 進歩（類 progress [4,053]，advancement [245]）/ 動 進行させる

□ **understand** [12,582]：動 理解する（類 appreciate [452]，know [34,957]，learn [1,427]）

□ **molecular mechanism** [4,233]：分子機構

□ **differentiation** [20,676]：名 分化

□ **process** [50,139]：名 過程（類 course [7,149]，advancement [245]）/ 動 処理する

□ **hopefully** [64]：副 うまくいけば，願わくば

□ **lead to ～** [31,651]：～につながる

□ **more** [67,205]：副 / 形 より～な（反 less [23,420]）

□ **effective** [14,834]：形 効果的な

□ **therapy for ～** [2,467]：～の治療

□ **lung cancer** [2,400]：肺癌

訳 分化過程の分子機構の理解における最近の進歩は，うまくいけば肺癌のより効果的な治療につながるであろう

207 This method has been used as a **tool for** localizing the effect of visual stimuli.

□ **method** [32,975]：名 方法

□ **used as ～** [4,361]：～として使われる

□ **tool** [6,214]：名 道具，手段（類 method [32,975]，instrument [1,524]，means [5,270]）

tool for ～ [2,465]：～のための道具

□ **localize** [15,589]：動 局在化する，局在する（類 locate [12,854]）

- ☐ **effect** [106,593]：名 影響，効果
- ☐ **visual** [8,105]：形 視覚の
- ☐ **stimulus** [11,878]：名（複 stimuli）刺激（類 stimulation [21,013]）

訳 この方法は，視覚刺激の影響を局在化するための道具として使われてきた

208 Fission yeast can serve as an excellent model **system for** genetic analysis of cell-polarity determination.

- ☐ **fission yeast** [786]：分裂酵母
- ☐ **serve as ～** [7,135]：～として役立つ，～として働く
- ☐ **excellent** [2,945]：形 優れた（類 good [4,633]）
- ☐ **system** [54,451]：名 システム，系
  - **system for ～** [2,458]：～のためのシステム
- ☐ **genetic** [28,358]：形 遺伝的な
- ☐ **analysis** [85,671]：名 解析，分析
- ☐ **cell polarity** [491]：細胞極性
- ☐ **determination** [4,378]：名 決定（類 decision [2,859]），定量

訳 分裂酵母は，細胞極性決定の遺伝的解析のための優れたモデルシステムとして役立ちうる

209 This method has great **potential for** identifying the pathways that are altered in response to the mutant protein.

- ☐ **method** [32,975]：名 方法
- ☐ **great** [2,183]：形 大きな
- ☐ **potential for ～** [2,201]：～のための潜在能
- ☐ **identify** [73,456]：動 同定する
- ☐ **pathway** [59,666]：名 経路
- ☐ **alter** [21,411]：動 変化させる
- ☐ **in response to ～** [16,447]：～に応答して
- ☐ **mutant** [79,727]：形 変異の / 名 変異体

訳 この方法は，変異タンパク質に応答して変化する経路を同定するための大きな潜在能を持つ

210 The aim of this study was to characterize **pathways for** the uptake and intraneuronal trafficking of protease-resistant prion protein.

- □ aim ［4,574］：名 目的 / 動 目的とする
- □ characterize ［27,658］：動 特徴づける（類 typify ［177］, clarify ［1,417］）
- □ pathway for ～ ［2,129］：～の経路
- □ uptake ［11,299］：名 取り込み
- □ intraneuronal ［119］：形 神経細胞内の
- □ trafficking ［4,296］：名 輸送（類 transport ［18,476］, translocation ［9,431］）
- □ protease ［10,907］：名 プロテアーゼ
- □ resistant ［12,791］：形 抵抗性の（類 tolerant ［921］, refractory ［1,992］）
- □ prion ［1,907］：名 プリオン

訳 この研究の目的は，プロテアーゼ抵抗性のプリオンタンパク質の取り込みと神経細胞内輸送の経路を特徴づけることであった

211 After **adjustment for** age and sex, there was no significant association between race and the risk of death from heart disease.

- □ adjustment for ～ ［1,932］：～に対する調整
  after adjustment for ～ ［1,359］：～に対して調整した後
- □ sex ［6,877］：名 性別，性
- □ significant ［43,571］：形 有意な，著しい，重要な
- □ association between ～ ［4,561］：～の間の関連
- □ race ［2,175］：名 人種
- □ risk ［37,608］：名 リスク，危険
- □ death ［26,658］：名 死
- □ heart disease ［2,426］：心疾患

訳 年齢と性別に対して調整した後，人種と心疾患による死のリスクの間に有意な関連はなかった

212 These data provide additional **support for** the hypothesis that this technique could elucidate the tissue of origin of metastatic lesions.

- □ **provide** [56,723]：動 提供する，与える
- □ **additional** [16,295]：形 付加的な
- □ **support for ～** [1,609]：～に対する支持
- □ **the hypothesis that ～** [7,746]：～という仮説
- □ **technique** [14,941]：名 技術，手法，テクニック（類 method [32,975]，procedure [8,622]）
- □ **elucidate** [5,037]：動 解明する（類 reveal [42,563]，clarify [1,417]，uncover [1,208]）
- □ **tissue** [45,858]：名 組織
- □ **origin** [7,942]：名 起源，由来
- □ **metastatic** [3,719]：形 転移性の
- □ **lesion** [18,776]：名 病変（部）/ 動 破壊する

訳 これらのデータは，この技術が転位性病変の起源の組織を解明しうるという仮説に対する付加的な支持を提供する

213 Together, these findings provide a plausible **explanation for** the severity of the autoimmune diseases in the mutant mice.

- □ **together** [17,449]：副 共に，一緒に
  Together, [3,624]：まとめると，（類 taken together [5,918]，collectively [2,430]，in summary [1,701]，in conclusion [2,471]）
- □ **finding** [35,256]：名 知見
- □ **provide** [56,723]：動 提供する，与える
- □ **plausible** [667]：形 もっともらしい（類 feasible [1,123]）
- □ **explanation** [2,329]：名 説明（類 interpretation [2,255]，description [1,540]）
  explanation for ～ [1,582]：～に対する説明
- □ **severity** [5,647]：名 重症度
- □ **autoimmune disease** [1,624]：自己免疫疾患
- □ **mutant** [79,727]：形 変異の / 名 変異体

訳 まとめると，これらの知見は変異マウスにおける自己免疫疾患の重症度に対するもっともらしい説明を提供する

214 Rising incidence of tuberculosis has prompted the **search for** alternative biomarkers using newly developed postgenomic approaches.

- □ **rise** [6,494]：動 上昇する（類 increase [146,670]）/ 名 上昇
- □ **incidence** [8,627]：名 発生率，頻度
- □ **tuberculosis** [5,173]：名 結核
- □ **prompt** [1,212]：動 駆り立てる，促す
- □ **search** [5,645]：名 探索（類 survey [3,519]，exploration [623]，investigation [6,218]，research [8,357]）/ 動 探索する

  search for ～ [1,455]：～の探索
- □ **alternative** [8,904]：形 代替の / 名 代替物
- □ **biomarker** [978]：名 生物マーカー
- □ **newly** [4,982]：副 新たに
- □ **develop** [38,325]：動 開発する，発症する，発達する
- □ **postgenomic** [34]：形 ポストゲノムの
- □ **approach** [21,009]：名 アプローチ，方法

訳 結核の発生率の上昇は，新たに開発されたポストゲノムのアプローチを使って代替の生物マーカーの探索を駆り立ててきた

215 This negative-strand RNA virus has inherent **specificity for** replication in tumor cells due to their attenuated antiviral responses.

- □ **negative-strand RNA** [115]：マイナス鎖RNA
- □ **virus** [47,464]：名 ウイルス
- □ **inherent** [1,094]：形 固有の（類 intrinsic [5,126]，endogenous [13,368]）
- □ **specificity for ～** [1,377]：～に対する特異性
- □ **replication** [21,221]：名 複製
- □ **tumor cell** [7,332]：腫瘍細胞
- □ **due to ～** [18,077]：～のせいで，～のゆえに

- □ **attenuate** [8,125]：動 減弱する（類 decrease [51,139], reduce [57,875], diminish [5,477]）
- □ **antiviral** [3,004]：形 抗ウイルス性の / 名 抗ウイルス剤
- □ **response** [104,667]：名 反応，応答

訳 このマイナス鎖RNAウイルスは，それらの減弱された抗ウイルス反応ゆえに腫瘍細胞における複製に対する固有の特異性を持っている

216 Using **data from** an observational study of 300 healthy adults, we analyzed the association of body mass index with dietary intake of carbohydrates.

- □ **data from ～** [3,852]：～からのデータ
- □ **observational** [586]：形 観察的な
- □ **healthy** [6,715]：形 健康な
- □ **analyze** [18,249]：動 分析する，解析する
- □ **association** [24,687]：名 関連（類 relationship [13,962], relation [4,419], relevance [1,908], correlation [9,923], link [28,890]），結合
- □ **body mass index** [1,523]：ボディマス指数，肥満度指数
- □ **dietary** [4,185]：形 食事性の
- □ **intake** [5,902]：名 摂取
- □ **carbohydrate** [3,117]：名 炭水化物

訳 300人の健康な成人の観察研究からのデータを使って，我々はボディマス指数と炭水化物の食事摂取との関連を解析した

217 In these mice, serotonin levels remained unchanged, and dopamine uptake and **release from** nerve terminals were normal.

- □ **serotonin** [2,541]：名 セロトニン
- □ **remain** [26,736]：動 ～のままである
- □ **unchanged** [2,413]：形 変化しない（類 unaltered [701], invariant [1,221]）
- □ **dopamine** [5,916]：名 ドパミン
- □ **uptake** [11,299]：名 取り込み
- □ **release** [23,508]：名 放出（類 discharge [2,812], secretion [12,758]）
  release from ～ [3,155]：～からの放出

□ nerve terminal [761]：神経終末

□ normal [44,556]：形 正常な

訳 これらのマウスにおいて，セロトニンのレベルは変化しないままであり，そしてドパミンの取り込みと神経終末からの放出は正常であった

218 Fecal **samples from** healthy children who received oral poliovirus vaccine were found to contain variants of Sabin vaccine viruses.

□ fecal [768]：形 糞便の

□ sample [22,785]：名 サンプル，試料（類 specimen [5,996]，preparation [5,341]）/ 動 試料採集する

sample from ～ [1,906]：～からのサンプル

□ healthy [6,715]：形 健康な

□ receive [19,081]：動 受ける（類 take [13,879]，undergo [19,421]）

□ oral [6,343]：形 経口の，口腔の

□ poliovirus [851]：名 ポリオウイルス

□ vaccine [10,085]：名 ワクチン

□ found to ～ [15,859]：～することが見つけられる

□ contain [71,403]：動 含む（類 include [59,168]，involve [49,378]）

□ variant [13,991]：名 変異体（類 mutant [79,727]）

□ Sabin vaccine [7]：セービンワクチン

□ virus [47,464]：名 ウイルス

訳 ポリオウイルスワクチンの経口接種を受けた健康な子供からの糞便サンプルは，セービンワクチンウイルスの変異体を含むことが見つけられた

219 We assessed **differences in** serum hyaluronan levels between the two groups, adjusting for ethnicity, sex, age, and body mass index.

□ assess [20,743]：動 評価する（類 evaluate [19,786]，estimate [13,499]）

□ difference in ～ [16,401]：～の違い

□ serum [21,260]：名 血清

□ hyaluronan [707]：名 ヒアルロン酸，ヒアルロナン

□ group [69,369]：名 群，グループ，基 / 動 グループ化する

□ **adjust** [5,867]：動 調整する（類 correct [4,898], control [78,085]）
adjusting for ～ [880]：～に対して調整する
□ **ethnicity** [708]：名 民族性
□ **body mass index** [1,523]：ボディマス指数，肥満度指数

訳 我々は，民族性，性別，年齢およびボディマス指数に対して調整して，その２群の間の血清ヒアルロン酸レベルの違いを評価した

220 A significant **reduction in** ejection fraction, as assessed by repeated echocardiography, was also observed.

□ **significant** [43,571]：形 有意な，著しい，重要な
□ **reduction in ～** [12,245]：～の低下
□ **ejection fraction** [1,207]：駆出率
□ **as assessed by ～** [1,146]：～によって評価されるように
□ **repeated** [3,341]：形 頻回の
□ **echocardiography** [1,469]：名 心エコー
□ **observe** [51,935]：動 観察する

訳 頻回の心エコーによって評価されたように，駆出率の有意な低下もまた観察された

221 Mobilization of endothelial progenitor cells with cytokines was found to enhance bone marrow cell **incorporation into** ischemic myocardium.

□ **mobilization** [1,747]：名 動員，可動化（類 recruitment [6,258]）
□ **endothelial** [17,212]：形 内皮の
□ **progenitor cell** [2,770]：前駆細胞
□ **cytokine** [19,780]：名 サイトカイン
□ **found to ～** [15,859]：～することが見つけられる
□ **enhance** [32,256]：動 増強する
□ **bone marrow cell** [789]：骨髄細胞
□ **incorporation** [4,734]：名 取り込み（類 uptake [11,299]）
incorporation into ～ [704]：～への取り込み
□ **ischemic** [4,608]：形 虚血性の

☐ **myocardium** [2,017]：名 心筋（類 cardiac muscle [549]）

訳 サイトカインによる内皮前駆細胞の動員は，虚血心筋への骨髄細胞の取り込みを増強することが見つけられた

222 **Data on** the association between androgen receptor polymorphisms and ovarian cancer are inconclusive.

☐ **data on ～** [2,130]：～に関するデータ

☐ **association between ～** [4,561]：～の間の関連

☐ **androgen receptor** [860]：アンドロゲン受容体

☐ **polymorphism** [6,809]：名 多型性

☐ **ovarian cancer** [1,780]：卵巣癌

☐ **inconclusive** [185]：形 決定的でない，不確定の（類 uncertain [1,163]，indefinite [113]）

訳 アンドロゲン受容体の多型性と卵巣癌の間の関連に関するデータは，決定的でない

223 The identification of numerous microtubule-associated proteins may have a profound **impact on** the study and treatment of human genetic disease.

☐ **identification** [10,858]：名 同定（類 detection [11,374]，discovery [3,618]）

☐ **numerous** [4,565]：形 多数の（類 many [26,628]）

☐ **microtubule-associated protein** [378]：微小管結合タンパク質

☐ **profound** [2,293]：形 著明な（類 striking [2,590]，marked [6,470]，prominent [2,934]，dramatic [3,523]，remarkable [1,492]），深刻な

☐ **impact** [6,642]：名 影響（類 effect [107,038]，influence [16,903]），衝撃（類 impulse [453]）/ 動 影響する

impact on ～ [1,925]：～に対する影響

☐ **treatment** [60,138]：名 治療，処理

☐ **genetic disease** [353]：遺伝性疾患

訳 多数のマイクロチューブ結合タンパク質の同定は，ヒトの遺伝性疾患の研究と治療に対する著明な影響をもつかもしれない

224 There have been few **studies on** the mechanisms that underlie the function of this novel isoform encoded by an alternatively spliced transcript.

- ☐ study on ～ [2,017]：～に関する研究
- ☐ **underlie** [10,696]：動 根底にある
- ☐ **function** [92,343]：名 機能 / 動 機能する
- ☐ **novel** [30,313]：形 新規の，新しい
- ☐ **isoform** [12,866]：名 アイソフォーム
- ☐ **encode** [37,774]：動 コードする
- ☐ **alternatively** [1,820]：副 オルターナティブに，代わりに（類 instead [4,422]）
- ☐ **splice** [12,147]：動 スプライシングする
- ☐ **transcript** [14,878]：名 転写物

訳 オルターナティブにスプライシングされた転写物によってコードされるこの新規のアイソフォームの機能の根底にある機構に関する研究はほとんどない

225 This study provides new **information on** the molecular basis of resistance and the evolution of resistance.

- ☐ **provide** [56,723]：動 提供する，与える
- ☐ information on ～ [1,758]：～に関する情報
- ☐ **molecular basis** [2,208]：分子基盤
- ☐ **resistance** [20,333]：名 抵抗性（類 tolerance [6,634]）
- ☐ **evolution** [7,907]：名 進化

訳 この研究は，抵抗性の分子基盤と抵抗性の進化に関する新しい情報を提供する

226 The major histocompatibility complex polymorphism has critical **influence on** the properties of the selected cytotoxic T lymphocyte repertoire.

- ☐ **major histocompatibility complex** [2,089]：主要組織適合遺伝子複合体
- ☐ **polymorphism** [6,809]：名 多型性
- ☐ **critical** [21,712]：動 決定的に重要な

☐ influence on ～ [1,402]：～に対する影響

☐ property [21,741]：名 性質

☐ select [10,123]：動 選択する（類 choose [1,765], screen [14,896]）

☐ cytotoxic T-lymphocyte [2,770]：細胞傷害性Tリンパ球

☐ repertoire [2,306]：名 レパートリー

訳 主要組織適合遺伝子複合体の多型性は，選択された細胞傷害性Tリンパ球レパートリーの性質に対する決定的な影響を持つ

227 The enzyme displayed a strong **dependence on** the ionic strength of the buffer.

☐ enzyme [49,164]：名 酵素

☐ display [15,251]：動 示す，提示する（類 exhibit [27,339], indicate [73,855], show [148,875]）/ 名 提示

☐ dependence [6,628]：名 依存性

dependence on ～ [955]：～への依存性

☐ ionic strength [1,104]：イオン強度

☐ buffer [3,563]：名 緩衝液 / 動 緩衝する

訳 その酵素は，緩衝液のイオン強度への強い依存性を示した

228 This transient loss of tetramer binding is associated with reduced **signaling through** the T-cell receptor.

☐ transient [9,961]：形 一過性の

☐ loss [28,751]：名 喪失，減少

☐ tetramer [2,303]：名 四量体，テトラマー

☐ binding [133,851]：名 結合

☐ associated with ～ [54,439]：～と関連する（類 related to [10,949], correlated with [9,967], linked to [5,539]）

☐ reduced [41,679]：低下した

☐ signaling through ～ [1,145]：～を経るシグナル伝達

☐ T-cell receptor [2,229]：T細胞受容体

訳 四量体結合のこの一過性の喪失は，T細胞受容体を経る低下したシグナル伝達と関連する

229 Chronic bronchitis is associated with impaired **resistance to** bronchial infection.

- ☐ chronic bronchitis [59]：慢性気管支炎
- ☐ associated with ～ [54,439]：～と関連する
- ☐ impair [10,580]：動 損なう，障害する（類 compromise [2,557]，damage [15,696]，lesion [18,776]，disrupt [7,792]）
- ☐ **resistance to ～** [5,635]：～に対する抵抗性
- ☐ bronchial [777]：形 気管支の
- ☐ infection [44,601]：名 感染

訳 慢性気管支炎は，気管支の感染に対する損なわれた抵抗性と関連する

230 Accumulating data emphasize the role of genetic factors as a cause of increased **susceptibility to** adverse drug responses.

- ☐ accumulate [6,498]：動 蓄積する（類 deposit [1,964]）
- ☐ emphasize [1,584]：動 強調する（類 underscore [1,128]，highlight [2,970]）
- ☐ role [87,150]：名 役割
- ☐ genetic factor [573]：遺伝因子
- ☐ cause [46,697]：名 原因 / 動 引き起こす
- ☐ increased [73,539]：増大した
- ☐ **susceptibility to ～** [3,082]：～に対する感受性
- ☐ adverse [4,131]：形 有害な（類 deleterious [1,383]，detrimental [613]，harmful [321]）
- ☐ drug response [105]：薬物応答

訳 蓄積するデータは，有害な薬物応答に対する増大した感受性の原因として遺伝的な因子の役割を強調する

231 Type 2 diabetes occurring spontaneously in rhesus monkeys shows a striking **similarity to** human diabetes in clinical features.

- ☐ diabetes [10,235]：名 糖尿病
- ☐ occur [42,905]：動 起こる，生じる

□ **spontaneously** [1,787]：副 自然に，自発的に（類 naturally [2,787]）

□ **rhesus monkey** [796]：アカゲザル

□ **show** [148,875]：動 示す

□ **striking** [2,590]：形 著しい（類 marked [6,470]，prominent [2,934]，significant [43,571]）

□ **similarity** [7,722]：名 類似性（類 homology [8,736]，analogy [393]）

similarity to ～ [2,596]：～に対する類似性

□ **clinical** [31,256]：形 臨床の

□ **feature** [15,065]：名 特徴

訳 アカゲザルにおいて自然に起こる2型糖尿病は，臨床的な特徴においてヒトの糖尿病に対する著しい類似性を示す

232 RNA-binding motif is known to have **homology to** RNA-binding proteins, but its function remains largely unknown.

□ **RNA-binding** [2,280]：RNA 結合

□ **motif** [18,066]：名 モチーフ

□ **known to ～** [8,375]：～すると知られている

□ **homology** [8,736]：名 相同性，ホモロジー（類 similarity [7,722]）

homology to ～ [2,496]：～に対する相同性

□ **function** [92,343]：名 機能 / 動 機能する

□ **remain** [26,736]：動 ～のままである

□ **largely** [6,410]：副 大部分は，（否定文で）ほとんど

□ **unknown** [11,802]：形 知られていない，未知の

訳 RNA結合モチーフは，RNA結合タンパク質への相同性を持つことが知られているが，その機能はほとんど知られていないままである

233 **Treatment with** this monoclonal antibody alone had relatively little effect on survivin and apoptosis.

□ treatment with ～ [8,802]：～による処理，～による治療

□ **monoclonal antibody** [5,131]：単クローン抗体，モノクローナル抗体

□ **alone** [13,764]：副 単独で

- □ **relatively** [9,096]：副 比較的
- □ **little** [12,531]：形 小さい / 副形 ほとんど～ない
- □ **effect on ～** [23,839]：～に対する影響
- □ **survivin** [777]：名 サバイビン
- □ **apoptosis** [31,507]：名 アポトーシス

訳 このモノクローナル抗体のみによる処理は，サバイビンとアポトーシスに対して比較的小さい影響しかもたなかった

234 When red cells are damaged, hemoglobin is liberated into the circulation and forms a **complex with** circulating haptoglobin.

- □ **red cell** [646]：赤血球
- □ **damage** [15,696]：動 損傷する（類 injure [1,498]，impair [10,580]）/ 名 障害，損傷
- □ **hemoglobin** [2,855]：名 ヘモグロビン
- □ **liberate** [345]：動 遊離する（類 release [29,830]）
- □ **circulation** [1,854]：名 循環
- □ **form** [67,531]：動 形成する（類 constitute [3,343]）/ 名 型
- □ **complex with ～** [5,345]：～との複合体
- □ **circulate** [3,729]：動 循環する
- □ **haptoglobin** [136]：名 ハプトグロビン

訳 赤血球が損傷されると，ヘモグロビンが循環に遊離され，そして循環ハプトグロビンとの複合体を形成する

2 名詞＋ that（同格の that）の文例

文例No			用例数
395	evidence that ～	～という証拠	9,968
177	the hypothesis that ～	～という仮説	7,746
105	the possibility that ～	～という可能性	3,349
317	the fact that ～	～という事実	2,097
235	observation that ～	～という観察	1,858
236	the idea that ～	～という考え	1,262
237	the notion that ～	～という考え	1,073
238	the view that ～	～という見解	702

235 This suggestion is based on the **observation that** at least 20 % of sporadic melanomas arise in association with atypical cells.

- □ **suggestion** [838]：名 提案，示唆（類 proposal [870]）
- □ **based on ～** [19,105]：～に基づいた
- □ **observation that ～** [1,858]：～という観察
- □ **at least** [16,456]：少なくとも
- □ **sporadic** [1,847]：形 孤発性の，散在性の
- □ **melanoma** [5,108]：名 メラノーマ，黒色腫
- □ **arise** [6,458]：動 生じる，起こる
- □ **in association with ～** [1,262]：～と関連して
- □ **atypical** [1,421]：形 異型の，非定型の

訳 この提案は，孤発性メラノーマの少なくとも20％が異型細胞と関連して生じるという観察に基づいている

236 These findings support **the idea that** high iodine intake can induce autoimmune thyroiditis in genetically predisposed animals.

- □ **finding** [35,256]：名 知見
- □ **support** [28,621]：動 支持する / 名 支持
- □ **the idea that ～** [1,262]：～という考え（類 the notion that [1,073], the view that [702]）
- □ **iodine** [392]：名 ヨウ素
- □ **intake** [5,902]：名 摂取
- □ **induce** [116,671]：動 誘導する
- □ **autoimmune** [4,490]：形 自己免疫性の
- □ **thyroiditis** [100]：名 甲状腺炎
- □ **genetically** [4,530]：副 遺伝的に
- □ **predispose** [1,375]：動 素因になる，病気に罹らせる

訳 これらの知見は，高いヨードの摂取が遺伝的に素因のある動物において自己免疫性甲状腺炎を誘導しうるという考えを支持する

237 Of some concern is **the notion that** obstructive sleep apnea may result in sudden death during sleep.

- □ **concern** [3,254]：名 重大事，懸念，関心 / 動 関係させる
- □ **the notion that ～** [1,073]：～という考え（類 the idea that [1,262], the view that [702]）
- □ **obstructive sleep apnea** [152]：閉塞型睡眠時無呼吸
- □ **result in ～** [48,455]：～という結果になる
- □ **sudden** [1,241]：形 突然の（類 abrupt [445]）
- □ **death** [26,658]：名 死
- □ **sleep** [3,654]：名 睡眠 / 動 眠る

訳 閉塞型睡眠時無呼吸が睡眠の間の突然死という結果になるかもしれないという考えは，いくぶん重大である

238 These results are compatible with **the view that** metabolic environment per se causes complications independent of genetic factors.

☐ **compatible with ～** [987]：～と矛盾しない，～に適合する

☐ **the view that ～** [702]：～という見解（類 the idea that [1,262]，the notion that [1,073]）

☐ **metabolic** [7,187]：形 代謝性の

☐ **environment** [7,753]：名 環境

☐ **per se** [720]：副 それ自体が（で）

☐ **cause** [46,697]：動 引き起こす / 名 原因

☐ **complication** [5,903]：名 合併症

☐ **independent of ～** [7,681]：～と関係ない，～に依存しない

☐ **genetic factor** [573]：遺伝因子

訳 これらの結果は，代謝性の環境それ自体は遺伝的因子と関係ない合併症を引き起こすという見解と矛盾しない

V. つなぎの表現

文をつなぐ表現は，文章を組立てるうえで非常に重要である．よく使われるつなぎの表現としては，**副詞／副詞的熟語**，**接続詞**，**副詞句を導く熟語**などがある．

1 逆説の文例

A. 副詞／副詞的熟語

逆説の意味の副詞／副詞的熟語は文頭で用いられることが多い．

文例No			用例数
239	However,	しかし，	45,056
240	Nevertheless,	にもかかわらず，	1,339
241	Nonetheless,	にもかかわらず，	548
242	Instead,	代わりに，	1,554
243	Alternatively,	代わりに，	571
152	Conversely,	逆に，	2,468
244	On the contrary,	逆に，それどころか，	121
245	In contrast,	対照的に，	19,491
246	By contrast,	対照的に，	2,160
247	On the other hand,	他方では，一方では，	1,308

239 **However,** the underlying mechanisms remain poorly understood.

- □ **However,** [45,056]：副 しかし，（類 but [136,113]）
- □ **underlying mechanism** [808]：根底にある機構
- □ **remain** [26,736]：動 ～のままである
- □ **poorly** [5,935]：副 あまり～でない
  poorly understood [3,015]：あまり理解されていない

訳 しかし，根底にある機構はあまり理解されていないままである

240 **Nevertheless,** some of the patients retained sufficient cognitive and verbal activity to perform mental status examinations at a normal level.

- ☐ **nevertheless** [1,753]：副 にもかかわらず（類 nonetheless [774]，however [18,707]）
  Nevertheless, [1,339]：にもかかわらず，
- ☐ **retain** [6,883]：動 保持する（類 hold [2,518]，keep [1,062]，maintain [13,333]，sustain [6,146]）
- ☐ **sufficient** [11,099]：形 十分な（類 adequate [1,554]）
- ☐ **cognitive** [3,746]：形 認知の
- ☐ **verbal** [464]：形 言語の
- ☐ **activity** [33,998]：名 活動，活性
- ☐ **perform** [21,488]：動 行う，実行する
- ☐ **mental** [2,424]：形 精神的な，知的な
- ☐ **status** [8,697]：名 状態
- ☐ **examination** [6,511]：名 検査
- ☐ **normal** [44,556]：形 正常な

訳 にもかかわらず，患者の幾人かは正常なレベルで精神状態検査を行う十分な認知および言語の活動を保持していた

241 **Nonetheless,** infection and rejection are the two major problems.

- ☐ **nonetheless** [774]：副 にもかかわらず（類 nevertheless [1,753]，however [18,707]）
  Nonetheless, [548]：にもかかわらず，
- ☐ **infection** [44,601]：名 感染
- ☐ **rejection** [7,059]：名 拒絶
- ☐ **major** [30,348]：形 主要な
- ☐ **problem** [6,340]：名 問題（類 matter [2,581]，issue [4,629]）

訳 にもかかわらず，感染と拒絶は２つの主要な問題である

V つなぎの表現

242 **Instead,** the interplay of various intracellular signaling pathways probably account for the functional synergy.

- □ **instead** [4,422]：副 代わりに（類 alternatively [1,820]）
  Instead, [1,554]：代わりに，
- □ **interplay** [787]：名 相互作用（類 interaction [63,464]）
- □ **various** [14,339]：形 さまざまな（類 a variety of [9,315]，diverse [6,826]）
- □ **intracellular** [20,566]：形 細胞内の
- □ **signaling pathway** [9,274]：シグナル経路
- □ **probably** [5,067]：副 おそらく，ありそうに（類 presumably [2,634]，conceivably [89]，likely [18,320]，maybe [37]，perhaps [2,517]，possibly [4,640]，potentially [6,206]）
- □ **account for ～** [8,259]：～を説明する，～を占める
- □ **functional** [34,575]：形 機能的な
- □ **synergy** [883]：名 相乗作用

訳 代わりに，さまざまな細胞内シグナル経路の相互作用はおそらくその機能的相乗作用を説明する

243 **Alternatively,** computer-based clinical support systems should be introduced that make it possible for physicians to utilize optimal antibiotic choices.

- □ Alternatively, [571]：代わりに，
- □ **computer-based** [110]：コンピュータを使った
- □ **clinical** [31,256]：形 臨床の
- □ **support** [28,621]：名 支援，支持（類 assistance [267]，aid [2,134]）／動 支持する
- □ **system** [54,451]：名 システム，系
- □ **introduce** [6,289]：動 導入する，紹介する
- □ **make it possible to ～** [443]：～することを可能にする
- □ **physician** [4,789]：名 内科医，医師
- □ **utilize** [6,566]：動 利用する（類 use [182,256]，exploit [1,885]，take advantage of [673]，employ [5,390]）
- □ **optimal** [4,845]：形 至適な，最適な
- □ **antibiotic** [2,279]：名 抗生剤／形 抗菌の

□ choice [2,317]：名 選択（類 selection [10,687]，option [1,433]）

訳 代わりに，内科医が最適な抗生剤選択を利用するのを可能にするコンピュータを使った臨床支援システムが導入されるべきである

244 **On the contrary,** no significant changes in cytokine levels were found between the patients and healthy volunteers.

□ On the contrary, [121]：副 逆に，それどころか，（類 conversely [2,765]）
□ significant [43,571]：形 有意な，著しい，重要な
□ change in ～ [30,981]：～の変化
□ cytokine [19,780]：名 サイトカイン
□ healthy [6,715]：形 健康な
□ volunteer [1,979]：名 ボランティア

訳 逆に，患者と健康なボランティアの間にサイトカインレベルの有意な変化は見つけられなかった

245 **In contrast,** anti-γδ antibody potently inhibited proliferation of γδ T cells.

□ In contrast, [19,491]：副 対照的に，（類 by contrast [2,428]）
□ antibody [36,724]：名 抗体
□ potently [1,443]：副 強力に（類 strongly [11,204]，robustly [260]，intensively [150]，intensely [271]）
□ inhibit [52,537]：動 抑制する（類 suppress [12,609]，repress [5,427]，block [31,876]，abrogate [3,332]，prevent [19,897]）
□ proliferation [19,120]：名 増殖

訳 対照的に，抗γδ抗体はγδT細胞の増殖を強力に抑制した

246 **By contrast,** there are no currently available vaccines for the bacterium responsible for syphilis.

□ By contrast, [2,160]：副 対照的に，（類 in contrast, [20,128]）
□ currently available [552]：現在利用できる
□ vaccine [10,085]：名 ワクチン
□ bacterium [11,218]：名（複 bacteria）細菌

☐ **responsible for ～** [12,456]：～に対して責任のある，～の原因である

☐ **syphilis** [215]：名 梅毒

訳 対照的に，梅毒の原因である細菌に対して現在有用なワクチンはない

247 **On the other hand,** some forms of cancer therapy have been reported to be very effective and well tolerated even with advanced age.

☐ **On the other hand,** [1,308]：他方では，一方では，（類 meanwhile [73]）

☐ **form** [9,556]：名 型／動 形成する

☐ **therapy** [28,037]：名 治療，療法（類 treatment [60,138]，care [11,307]，practice [3,583]）

☐ **reported to ～** [2,484]：～すると報告される

☐ **very** [14,515]：副 非常に，とても（類 unusually [935]，extremely [2,414]，greatly [5,065]，markedly [7,503]）

☐ **effective** [14,834]：形 効果的な

☐ **well** [45,774]：副 よく／形 よい

☐ **tolerated** [2,133]：形 耐容性を示した

☐ **advanced age** [113]：高齢

訳 他方では，いくつかの型の癌療法は非常に効果的で高齢でも耐容性がよいと報告されている

B. 副詞節を導く接続詞

文例No			用例数
248	although ～	～であるけれども，～にもかかわらず	38,035
－	though ～	～であるけれども，～にもかかわらず	3,282
249	even though ～	たとえ～であるにしても，～であるけれども	1,841
250	while ～	～であるけれども，～の間に，だが一方～	27,429
251	whereas ～	だが一方～	34,018

248 **Although** all communication between neurons is known to occur through synapses, little is known about the mechanisms inducing their formation.

- □ although [38,035]：接 ～であるけれども，にもかかわらず（類 though [3,282]，while [27,429]）
- □ communication [2,232]：名 コミュニケーション，連絡
- □ neuron [39,544]：名 ニューロン
- □ known to ～ [8,375]：～すると知られている
- □ occur through ～ [989]：～を経て起こる
- □ synapse [6,527]：名 シナプス
- □ little is known about ～ [3,195]：～についてはほとんど知られていない
- □ induce [116,671]：動 誘導する
- □ formation [41,324]：名 形成

訳 ニューロンの間のすべてのコミュニケーションはシナプスを経て起こることが知られているけれども，それらの形成を誘導する機構についてはほとんど知られていない

249 **Even though** the process of transdifferentiation has been well studied and established in amphibian systems, whether mammalian cells possess the same potential remains unclear.

- □ even though [1,841]：たとえ～であるにしても，～であるけれども（類 although [38,035]，even if [369]）
- □ process [50,139]：名 過程／動 処理する

☐ **transdifferentiation** [160]：名 分化転換

☐ **study** [167,914]：動 研究する（類 investigate [30,619], examine [39,969], survey [3,519], search [5,645], explore [6,658]）／名 研究

☐ **establish** [19,016]：動 確立する（類 confirm [17,664]）

☐ **amphibian** [379]：名 両生類

☐ **system** [54,451]：名 システム，系

☐ **mammalian** [15,396]：形 哺乳類の

☐ **same** [24,047]：形 同じ（類 identical [9,552]）

☐ **potential** [37,200]：名 潜在能（類 capability [1,731], ability [29,720], competence [1,025]），可能性／形 潜在的な，可能な

☐ **remain** [26,736]：動 ～のままである

☐ **unclear** [5,398]：形 不明な

訳 たとえ分化転換の過程は両生類系においてよく研究されそして確立されているにしても，哺乳類細胞が同じ潜在能を持つかどうかは不明なままである

250 **While** withdrawal symptoms are present in late stages of drug dependence, physicians should also be alert to the early behavioral signs of impaired social functioning.

☐ **while** [27,429]：接 ～であるけれども（類 although [38,035]），～の間に，だが一方（類 whereas [34,018]）

☐ **withdrawal symptoms** [45]：禁断症状

☐ **present** [41,922]：形 存在する，現在の／動 提示する

☐ **late stage** [796]：後期

☐ **drug dependence** [51]：薬物依存

☐ **physician** [4,789]：名 内科医，医師

☐ **also** [110,138]：副 ～もまた（類 too [1,210]）

☐ **alert** [290]：形 警戒する／動 警告する（類 warn [104]）

☐ **early** [31,638]：形 早期の／副 早く

☐ **behavioral** [4,542]：形 行動上の

☐ **sign** [2,742]：名 徴候（類 symptom [9,302], indication [1,386], manifestation [1,760]）

☐ **impair** [10,580]：動 障害する，損なう

☐ **social** [3,147]：形 社会的な

☐ functioning [2,224]：機能

訳 禁断症状は薬物依存の後期に存在するけれども，内科医はまた障害された社会生活機能の早期の行動上の徴候に警戒するべきである

251 Seventy-five percent of the examined human neurofibromas expressed progesterone receptor, **whereas** only 5% expressed estrogen receptor.

☐ percent [9,572]：名 パーセント，%
☐ examine [39,969]：動 調べる
☐ neurofibroma [101]：名 神経線維腫
☐ express [78,977]：動 発現する
☐ progesterone receptor [412]：プロゲステロン受容体
☐ whereas [34,018]：接 だが一方（類 while [27,429]）
☐ only [56,980]：副 たった〜だけ
☐ estrogen receptor [1,835]：エストロゲン受容体

訳 調べられたヒトの神経線維腫の75%がプロゲステロン受容体を発現していたが，一方，5%だけがエストロゲン受容体を発現していた

C. 副詞句を導く熟語／接続詞

文例No			用例数
252	in contrast to ～	～とは対照的に	6,868
253	as opposed to ～	～とは対照的に	611
254	contrary to ～	～とは反対に	882
255	despite ～	～にもかかわらず	12,361
256	in spite of ～	～にもかかわらず	444
257	Unlike ～	～と違って	2,716

252 **In contrast to** other translation initiation factors, these two factors colocalize to a specific cytoplasmic locus.

- □ **in contrast to ～** [6,868]：～とは対照的に（類 as opposed to [611]， contrary to [882]）
- □ **translation** [7,475]：名 翻訳
- □ **initiation** [10,562]：名 開始（類 onset [9,566]，start [6,641]）
- □ **factor** [103,476]：名 因子
- □ **colocalize** [2,673]：動 共存する（類 coexist [557]）
- □ **specific** [85,296]：形 特異的な
- □ **cytoplasmic** [12,439]：形 細胞質の
- □ **locus** [18,724]：名 部位，座位（類 region [77,257]，site [99,321]，area [19,443]）

訳 他の翻訳開始因子とは対照的に，これらの２つの因子は特異的な細胞質部位に共存する

253 The disagreement of the value with the results of other observers is explained in terms of individual metabolic properties **as opposed to** those of the larger population.

- □ **disagreement** [161]：名 不一致，相違（類 discordance [153]， inconsistency [206]，difference [33,598]，divergence [2,132]， disparity [956]）
- □ **value** [21,891]：名 値，価値
- □ **observer** [712]：名 観察者

☐ in terms of ～ [3,377]：～の点から

☐ explain [9,556]：動 説明する

☐ individual [24,178]：形 個々の，個々人の／名 個人

☐ metabolic [7,187]：形 代謝性の

☐ property [21,741]：名 性質

☐ as opposed to ～ [611]：～とは対照的に（類 in contrast to [6,868]，contrary to [882]）

☐ population [28,214]：名 集団，人口

訳 他の観察者の結果との値の不一致は，より大きな集団のそれらとは対照的に，個々人の代謝の性質の点から説明される

254 **Contrary to** the widely held assumption, rapid eye movement (REM) and non-REM sleep are not mutually exclusive states.

☐ contrary to ～ [882]：～とは反対に（類 in contrast to [6,868]，as opposed to [611]）

☐ widely [5,189]：副 広く

☐ hold [2,518]：動 保持する，支持する（類 retain [6,883]，keep [1,062]，maintain [13,333]，sustain [6,146]）

☐ assumption [1,542]：名 仮定，推定（類 hypothesis [13,999]，concept [2,660]，notion [1,569]，idea [2,167]，view [4,519]）

☐ rapid eye movement [191]：急速眼球運動（REM）

☐ REM sleep [430]：レム睡眠

☐ mutually [604]：副 相互に（類 reciprocally [286]）

☐ exclusive [883]：形 排他的な，独占的な

☐ state [32,696]：名 状態（類 status [8,697]，condition [30,859]，situation [1,437]）

訳 広く支持されている推定とは反対に，レム睡眠とノンレム睡眠は相互に排他的な状態ではない

255 **Despite** extensive data regarding the in vitro and in vivo activities of the drugs, early studies did not recognize that the high concentrations of these drugs used in vitro were unattainable in vivo.

- □ **despite ～** [12,361]：接 ～にもかかわらず（類 in spite of [444]）
- □ **extensive** [6,077]：形 広範な（類 wide [7,526], broad [4,339], widespread [2,588]）
- □ **regarding ～** [3,498]：前 ～に関する（類 concerning [1,266], with respect to [4,188], in relation to [1,190]）
- □ **in vitro** [41,334]：形 試験管内の／副 試験管内で
- □ **in vivo** [41,680]：形 生体内の／副 生体内で
- □ **activity** [33,998]：名 活性，活動
- □ **early** [31,638]：形 早期の／副 早く
- □ **recognize** [11,391]：動 認識する（類 appreciate [452], perceive [1,062], notice [95]）
- □ **concentration** [39,910]：名 濃度，集中
- □ **unattainable** [12]：形 達成できない

訳 その薬の試験管内および生体内活性に関する広範なデータにもかかわらず，早期の研究は試験管内で使われたこれらの薬の高い濃度は生体内では達成できないということに気づかなかった

256 **In spite of** extensive research on Parkinson's disease and Alzheimer's disease, preventive or long-term effective treatment strategies have not yet been proposed.

- □ **in spite of ～** [444]：～にもかかわらず（類 despite [12,361]）
- □ **extensive** [6,077]：形 広範な
- □ **research** [8,332]：名 研究，調査（類 study [167,914], test [45,998], examination [6,511], investigation [6,218], survey [3,519], work [12,304]）
- □ **treatment** [60,138]：名 治療，処理
- □ **Parkinson's disease** [1,529]：パーキンソン病
- □ **Alzheimer's disease** [2,946]：アルツハイマー病
- □ **preventive** [643]：形 予防の，予防的な（類 prophylactic [727]）
- □ **long-term** [10,326]：形 長期の

□ effective [14,834]：形 効果的な

□ strategy [13,405]：名 戦略（類 methodology [1,967]，method [32,975]）

□ propose [20,814]：動 提案する，提唱する

訳 パーキンソン病およびアルツハイマー病に関する広範な研究にもかかわらず，予防的あるいは長期の効果的な治療戦略はまだ提案されていない

257 Unlike renal disease, no useful laboratory test is available to make any general recommendations for drug dose adjustments in liver disease.

□ Unlike [2,716]：前 ～と違って（類 however [64,020]）

□ renal disease [833]：腎疾患

□ useful [9,391]：形 有用な，役に立つ（類 available [10,780]，helpful [430]）

□ laboratory test [114]：臨床検査（類 clinical examination [184]）

□ available [10,780]：形 利用できる（類 accessible [1,926]），入手できる

□ general [10,998]：形 一般的な（類 common [19,789]，normal [44,556]，usual [1,028]）

□ recommendation [1,296]：名 推奨，推薦（類 advice [141]）

□ drug dose [49]：薬剤投与量

□ adjustment [2,964]：名 調整

□ liver disease [1,302]：肝疾患

訳 腎疾患と違って，肝疾患における薬剤投与量調整に対する一般的な推奨をするために利用できる有用な臨床検査はない

2 肯定の文例

A. 副詞／副詞的熟語

文例No			用例数
258	therefore	それゆえ	17,502
259	hence	それゆえ	2,895
260	Accordingly,	したがって，それゆえ，	1,101
154	Consequently,	したがって，結果的に，	1,197
261	Thus,	このように，したがって，	24,291
262	then	それから	13,597
263	thereby	それによって	6,038
264	Furthermore,	さらに，	20,154
265	Further,	さらに，	2,307
266	Moreover,	さらに，	10,944
149	Additionally,	そのうえ，さらに，	3,649
267	In addition,	そのうえ，加えて，	19,413
268	indeed	実際に	2,437
269	in fact	実際に	1,140
270	similarly	同様に	5,807
271	likewise	同様に	1,094

258 **Therefore**, the physician should always take into consideration the possibility of an adverse drug response in the differential diagnosis.

- □ **therefore** [17,502]：副 それゆえ（類 hence [2,895], accordingly [1,249], consequently [2,153], thus [38,689]）
- □ **physician** [4,789]：名 内科医，医師
- □ **always** [1,581]：副 常に（類 constantly [190], routinely [735]）
- □ **take into consideration** [97]：考慮する（類 consider [8,440]）
- □ **possibility** [6,450]：名 可能性

□ adverse [4,131]：形 有害な

□ drug response [105]：薬物応答

□ differential diagnosis [170]：鑑別診断

訳 それゆえ，内科医は鑑別診断において薬物副作用の可能性を常に考慮すべきである

259 Hence, the most intense monitoring is needed in the first 6 to 8 weeks after transplantation.

□ hence [2,895]：副 それゆえ（類 therefore [17,502], accordingly [1,249], consequently [2,153], thus [38,689]）

□ most [44,482]：副 最も

□ intense [1,624]：形 強烈な，強力な，集中した（類 strong [11,381], robust [3,181], potent [11,330], powerful [2,820], intensive [2,341]）

□ monitoring [123]：名 モニタリング，監視（類 surveillance [1,894], guard [409], scrutiny [123]）

□ need [12,039]：動 必要とする / 名 必要性

□ transplantation [11,920]：名 移植（類 implantation [1,664], transplant [10,234], implant [3,147], explant [1,632], graft [11,730]）

訳 それゆえ，最も集中したモニタリングが移植後の最初の６から８週の間に必要とされる

260 Accordingly, only a few antivirals for respiratory viral illnesses are approved for use, each with very limited indications.

□ Accordingly, [1,101]：副 したがって，それゆえ，（類 consequently [2,153], therefore [17,502], hence [2,895], thus [38,689]）

□ a few [114]：わずかな，少しの（類 slight [983], subtle [1,183], modest [2,515]）

□ antiviral [3,004]：名 抗ウイルス剤 / 形 抗ウイルス性の

□ respiratory [6,788]：形 呼吸の，呼吸器の

□ viral [21,817]：形 ウイルスの，ウイルス性の

□ illness [3,106]：名 疾病，病気（類 disease [72,082], sickness [149]）

□ approve [813]：動 承認する（類 accept [1,866]）

□ use [182,256]：名 使用 / 動 使う

□ **each** [37,237]：名 おのおの，それぞれ（類 individual [24,178]）

□ **limited** [10,640]：形 限られた

□ **indication** [1,386]：名 適応症，徴候，指示（類 evidence [35,499]，demonstration [2,008]，representation [2,171]，symptom [9,302]，sign [2,848]）

訳 したがって，呼吸器のウイルス性疾病に対してはごく小数の抗ウイルス薬のみが使用することを承認されており，それぞれが非常に限られた適応症を持つ

261 **Thus,** the increased risk of dying of pneumonia is probably related to the decreased ability of the aged immune system to combat the pathogens.

□ **Thus,** [24,291]：副 このように，したがって，（類 accordingly [1,249]，consequently [2,153]，therefore [17,502]，hence [2,895]）

□ **die of ～** [571]：～で死ぬ

□ **risk** [37,608]：名 リスク，危険

□ **infected** [20,742]：感染した

□ **pneumonia** [1,737]：名 肺炎

□ **probably** [5,067]：副 おそらく，ありそうに

□ **relate** [38,355]：動 関連づける

related to ～ [10,949]：～に関連する（類 associated with [54,439]，correlated with [9,967]，linked to [5,539]）

□ **decreased** [26,518]：低下した

□ **ability** [29,720]：名 能力

□ **aged** [4,570]：形 老化した，高齢の（類 elderly [2,100]）

□ **immune system** [2,453]：免疫系

□ **combat** [311]：動 戦う / 名 戦い（類 fight [153]）

□ **pathogen** [7,954]：名 病原体

訳 このように，肺炎で死ぬ増大したリスクは老化した免疫系の病原体と戦う低下した能力とおそらく関連している

262 Those infected adults were selected and **then** screened for drug-selected resistance mutations and phylogenetic subtype.

☐ **infected** [20,742]：感染した

☐ **select** [10,123]：動 選択する

☐ **then** [13,597]：副 それから（類 thereby [6,038]）

☐ **screen** [14,896]：動 選別する，スクリーニングする / 名 スクリーン（類 sort [2,905]，filter [1,916]）

☐ **resistance** [20,333]：名 抵抗性

☐ **mutation** [70,231]：名 変異

☐ **phylogenetic** [2,894]：形 系統学的な

☐ **subtype** [5,629]：名 サブタイプ

訳 それらの感染した大人が選択され，それから薬剤選択性抵抗性変異と系統学的なサブタイプで選別された

263 The aim of the drug therapy was to induce vascular smooth muscle relaxation, **thereby** relieving spasm, raising resting blood flow, and limiting the degree of ischemia during attacks.

☐ **aim** [4,574]：名 目的 / 動 目的とする

☐ **drug therapy** [363]：薬物療法

☐ **induce** [116,671]：動 誘導する

☐ **vascular** [14,312]：形 血管の

☐ **smooth muscle** [5,630]：平滑筋

☐ **relaxation** [3,447]：名 弛緩，緩和

☐ **thereby** [6,038]：副 それによって（類 then [13,597]）

☐ **relieve** [999]：動 軽減する（類 alleviate [640]），解放する

☐ **spasm** [87]：名 けいれん（類 seizure [2,853]）

☐ **raise** [5,926]：動 上げる（類 elevate [13,039]，increase [146,670]）

☐ **resting** [3,509]：形 安静時の

☐ **blood flow** [3,412]：血流量

☐ **limit** [21,272]：動 限定する，制限する / 名 限界

□ **degree** [21,891]：名 程度（類 extent [9,276], grade [6,094]），度

□ **ischemia** [6,280]：名 虚血

□ **attack** [2,115]：名 発作（類 stroke [4,593], seizure [2,853]），攻撃

訳 その薬物療法の目的は，血管平滑筋弛緩を誘導し，それによってけいれんを軽減し，安静時血流量を上げ，そして発作の間の虚血の程度を限定することであった

264 **Furthermore,** our data suggest that genital epithelial cells may provide a barrier to HIV infection.

□ **furthermore,** [20,702]：副 さらに，（類 further [29,238], additionally [4,298], moreover [11,306]）

□ **suggest that 〜** [96,112]：〜ということを示唆する

□ **genital** [1,036]：形 生殖器の

□ **epithelial cell** [10,220]：上皮細胞

□ **provide** [56,723]：動 提供する，与える

□ **barrier** [5,355]：名 障害，障壁

□ **HIV** [29,027]：名 ヒト免疫不全ウイルス

□ **infection** [44,601]：名 感染

訳 さらに，我々のデータは生殖器上皮細胞がヒト免疫不全ウイルス感染への障壁を提供するかもしれないということを示唆する

265 **Further,** murine embryonic fibroblasts obtained from these mice did not die in response to oxygen deprivation.

□ **Further,** [2,307]：副 さらに（類 furthermore [20,702], additionally [4,298], moreover [11,306]）/ 形 さらに進んだ

□ **murine** [13,104]：形 マウスの

□ **embryonic** [10,071]：形 胚性の

□ **fibroblast** [12,722]：名 線維芽細胞

□ **obtained from 〜** [4,871]：〜から得られる

□ **die** [4,831]：動 死ぬ

□ **in response to 〜** [16,447]：〜に応答して

□ **oxygen** [10,233]：名 酸素

☐ deprivation [1,977]：名 欠乏（類 deficiency [6,979], depletion [6,011], lack [23,888]）

訳 さらに，これらのマウスから得られたマウス胚線維芽細胞は酸素欠乏に応答して死ぬことはかった

266 **Moreover**, the combination therapy was more effective than temozolomide treatment alone.

☐ Moreover, [11,306]：副 さらに，（類 furthermore [20,702], additionally [4,298]）

☐ combination [15,301]：名 併用，組合わせ（類 conjunction [2,071]）

☐ combination therapy [604]：併用療法

☐ effective [14,834]：形 効果的な

☐ temozolomide [140]：名 テモゾロミド

☐ treatment [60,138]：名 治療，処理

訳 さらに，併用療法はテモゾロミド療法単独よりも効果的であった

267 **In addition,** immunosuppressive therapy in animal trials has been reported to increase early mortality.

☐ In addition, [19,413]：そのうえ，加えて（類 additionally [4,298]）

☐ immunosuppressive therapy [286]：免疫抑制療法

☐ trial [13,519]：名 治験，試行

☐ reported to ～ [2,484]：～すると報告される

☐ increase [146,670]：動 増大させる，増大する / 名 増大

☐ early [31,638]：形 初期の，早期の / 副 早く

☐ mortality [12,864]：名 死亡率

訳 そのうえ，動物治験における免疫抑制療法は初期の死亡率を増大させることが報告されている

268 **Indeed**, nonmammalian nervous systems provide ideal platforms for the study of fundamental problems in neuroscience.

☐ indeed [2,437]：副 実際に（類 in fact [1,140]）

V つなぎの表現

- □ **nonmammalian** [60]：形 非哺乳類の
- □ **nervous system** [6,242]：神経系
- □ **provide** [56,723]：動 提供する，与える
- □ **ideal** [1,004]：形 理想の / 名 理想
- □ **platform** [1,178]：名 プラットホーム
- □ **fundamental** [3,010]：形 基礎的な，基本的な（類 basal [10,481]，basic [6,872]，necessary [12,049]，essential [23,051]）
- □ **problem** [6,340]：名 問題（類 issue [4,561]）
- □ **neuroscience** [241]：名 神経科学

訳 実際に，非哺乳類の神経系は神経科学における基礎的な問題の研究のための理想のプラットホームを提供する

269 **In fact**, these drugs may be harmful under certain circumstances.

- □ **in fact** [1,140]：副 実際に（類 indeed [2,437]）
- □ **harmful** [321]：形 有害な（類 adverse [4,131]，deleterious [1,383]，hazardous [79]）
- □ **certain** [7,346]：形 ある（類 some [26,452]），確かな
- □ **circumstance** [761]：名 環境，状況（類 environment [7,753]，case [25,633]，occasion [363]，situation [1,437]，condition [30,859]）

訳 実際に，これらの薬剤はある環境下で有害であるかもしれない

270 **Similarly**, ineffective helper T cell activation inhibits the immune response to blood group antigens in ABO-mismatched allograft recipients.

- □ **similarly** [5,807]：副 同様に（類 likewise [1,094]）
- □ **ineffective** [1,091]：形 無効な，無力な
- □ **helper T cell** [177]：ヘルパーT細胞
- □ **activation** [82,818]：名 活性化
- □ **inhibit** [52,537]：動 抑制する
- □ **immune response** [8,912]：免疫応答
- □ **blood group antigen** [71]：血液型抗原
- □ **mismatched** [1,100]：不適合の，不適正な（類 inappropriate [957]）

□ allograft [5,871]：名 同種移植

□ recipient [10,709]：名 レシピエント，移植患者

訳 同様に，無力なヘルパーT細胞の活性化は免疫応答を抑制するABO不適合な同種移植レシピエントにおける血液型抗原に対する免疫応答を抑制する

271 **Likewise**, the physical examination needs to be complete, with more emphasis placed on carefully measuring the blood pressure.

□ likewise [1,094]：副 同様に（類 similarly [5,807]）

□ physical examination [292]：理学的検査

□ need to ～ [4,886]：～する必要がある

□ complete [14,902]：形 完全な，徹底的な（類 full [12,117]，perfect [519]）/ 動 完了する

□ emphasis [844]：名 強調

□ place [7,331]：動 置く（類 lie [2,783]，put [394]）/ 名 場所

□ carefully [533]：副 注意深く

□ measure [34,591]：動 測定する

□ blood pressure [3,723]：血圧

訳 同様に，理学的検査は，血圧を注意深く測定することに対してより強調を置いて，完全である必要がある

B. 副詞節を導く接続詞／熟語

文例No			用例数
272	because ～	～なので	28,947
273	since ～	～なので，～以来	11,101
361	as ～	～なので，～につれて，～であるとき，～のように	267,565
274	so that ～	その結果～，それで～，～するように	1,839

272 **Because** ischemia can evoke the generation of reactive oxygen species, we explored the effect of oxidative stress on membrane phospholipid.

- □ **because** [28,947]：接 ～なので（類 since [11,101]，as [267,565]）
- □ **ischemia** [6,280]：名 虚血
- □ **evoke** [5,622]：動 誘起する，引き起こす（類 elicit [7,731]，provoke [549]，induce [116,671]）
- □ **generation** [10,962]：名 産生（類 production [27,611]），生成，世代
- □ **reactive oxygen species** [1,651]：活性酸素種
- □ **explore** [6,658]：動 探索する（類 study [167,914]，investigate [30,619]，examine [39,969]，survey [3,519]，search [5,645]）
- □ **effect of … on ～** [6,000]：…の～に対する影響
- □ **oxidative** [6,890]：形 酸化的な
- □ **stress** [18,967]：名 ストレス
- □ **membrane** [56,085]：名 膜
- □ **phospholipid** [4,448]：名 リン脂質

訳 虚血は活性酸素種の産生を誘起しうるので，我々は膜のリン脂質に対する酸化ストレスの影響を探索した

273 **Since** the cause of this disease is unknown and no cure is available, the physician must distinguish as early as possible the attendant risks for each patient.

- □ **since** [11,101]：接 ～なので（類 because [28,947]，as [267,565]），～以来
- □ **cause** [46,697]：名 原因 / 動 引き起こす
- □ **disease** [71,437]：名 疾患

☐ **unknown** [11,802]：形 知られていない，未知の

☐ **cure** [1,220]：名 治療（類 therapy [28,037], treatment [60,138], care [11,136]）/ 動 治療する

☐ **available** [10,780]：形 利用できる，入手できる

☐ **physician** [4,789]：名 内科医，医師

☐ **distinguish** [4,689]：動 識別する，区別する（類 discriminate [1,770]，discern [321]，sort [2,905]，differentiate [7,522]）

☐ **as early as possible** [7]：できるだけ早く

☐ **attendant** [110]：形 付帯する，付随する

☐ **risk for ～** [4,282]：～のリスク

訳 この疾患の原因は知られておらず利用できる治療法がないので，内科医はおのおのの患者の付帯リスクをできるだけ早く識別しなければならない

274 The doctor must work to develop a partnership with the patient **so that** he can identify and use diagnostic tests and treatments that are acceptable to the patient.

☐ **doctor** [207]：名 医師（類 physician [4,789]）

☐ **work** [12,304]：動 働く，努力する / 名 研究，仕事（類 study [167,914]，investigation [6,218]，research [8,330]）

☐ **develop** [38,325]：動 作りあげる，開発する，発症する，発達する

☐ **partnership** [127]：名 協力（類 cooperation [712]，collaboration [334]）

☐ **so that ～** [1,839]：その結果～，それで～，～するように

☐ **identify** [73,456]：動 同定する

☐ **diagnostic test** [334]：診断検査

☐ **treatment** [60,138]：名 治療，処理

☐ **acceptable** [758]：形 容認できる，受け入れられる（類 tolerable [136]）

訳 医師は，患者に受け入れられる診断テストと治療を同定し利用できるように患者との協力関係を作りあげるように努力しなければならない

C. 副詞句を導く熟語

文例No			用例数
275	in agreement with ～	～に一致して	1,140
276	in accordance with ～	～に一致して	319
277	, coincident with ～	～に一致して	833
278	according to ～	～に従って，～によれば	3,559
279	in addition to ～	～に加えて	7,445
280	because of ～	～のせいで，～ゆえに	7,878
281	Due to ～	～のせいで，～ゆえに	461
282	owing to ～	～のせいで，～ゆえに	782

275 **In agreement with** these findings, knockdown of Wnt signaling resulted in increased cell proliferation.

- ☐ in agreement with ～ [1,140]：～に一致して（類 coincident with [833]）
- ☐ **finding** [35,256]：名 知見
- ☐ **knockdown** [826]：名 ノックダウン
- ☐ **signaling** [42,594]：名 シグナル伝達
- ☐ **result in ～** [48,455]：～という結果になる
- ☐ **increased** [73,539]：増大した
- ☐ **cell proliferation** [6,197]：細胞増殖

訳 これらの知見に一致して，Wntシグナル伝達のノックダウンは増大した細胞増殖という結果になった

276 **In accordance with** these data, plasma corticosterone levels were decreased.

- ☐ in accordance with [319]：～に一致して（類 coincident with [833]）
- ☐ **plasma** [22,942]：名 血漿
- ☐ **corticosterone** [593]：名 コルチコステロン
- ☐ **decrease** [51,139]：動 低下させる（類 reduce [57,875]，diminish [5,477]，lower [25,400]，down-regulate [2,897]），低下する / 名 低下，減少

訳 これらのデータに一致して，血漿コルチコステロンレベルが低下した

277 Sleepiness peak typically occurs during the latter half of the night, **coincident with** maximal sleep drive within the brain.

- □ sleepiness〔130〕：名 眠気
- □ peak〔9,497〕：名 ピーク / 動 ピークになる
- □ typically〔2,935〕：副 主として（類 mostly〔1,734〕，primarily〔8,336〕，largely〔6,410〕），典型的に
- □ occur during ～〔2,005〕：～の間に起こる
- □ latter half〔24〕：後半
- □ coincident with ～〔833〕：～に一致して
- □ maximal〔5,405〕：形 最大の（類 largest〔2,072〕）
- □ sleep〔3,448〕：名 睡眠
- □ drive〔781〕：名 動因，駆動 / 動 駆動する
- □ brain〔28,011〕：名 脳

訳 脳内の最大の睡眠動因に一致して，眠気のピークは主として夜の後半の間に起こる

278 **According to** this mechanism, the enzyme undergoes multiple conformation changes during the catalytic cycle.

- □ according to ～〔3,559〕：～に従って，～によれば
- □ enzyme〔49,164〕：名 酵素
- □ undergo〔19,420〕：動 起こす，経験する
- □ multiple〔26,071〕：形 複数の
- □ conformation〔11,113〕：名 構造（類 structure〔65,777〕）
- □ change〔63,291〕：名 変化（類 alteration〔9,317〕，variation〔10,447〕，shift〔12,913〕）/ 動 変化する，変化させる
- □ catalytic〔13,686〕：形 触媒的な
- □ cycle〔23,639〕：名 サイクル，周期

訳 この機構にしたがって，その酵素は触媒的サイクルの間に複数の構造変化を起こす

279 **In addition to** abnormalities of muscle protein metabolism, heritable factors may contribute to susceptibility to external triggers for cardiac dysfunction.

- □ **in addition to ～** [7,445]：～に加えて
- □ **abnormality** [7,645]：名 異常
- □ **muscle** [25,413]：名 筋肉
- □ **metabolism** [8,783]：名 代謝
- □ **heritable** [571]：形 遺伝的な（類 genetic [28,358]）
- □ **factor** [103,476]：名 因子
- □ **contribute to ～** [19,994]：～に寄与する
- □ **susceptibility to ～** [3,082]：～に対する感受性
- □ **external** [3,834]：形 外部の
- □ **trigger** [8,460]：名 引き金，動 引き金を引く
- □ **cardiac** [17,231]：形 心臓の
- □ **dysfunction** [6,236]：名 機能障害

訳 筋肉タンパク質代謝の異常に加えて，遺伝的因子が心機能不全の外的引き金に対する感受性に寄与するかもしれない

280 **Because of** the biotechnology revolution, many novel vaccines will be developed and become available in the future.

- □ **because of ～** [7,878]：～ゆえに，～のせいで（類 due to [18,077]，owing to [782]）
- □ **biotechnology** [351]：名 バイオテクノロジー
- □ **revolution** [135]：名 革命，回転
- □ **novel** [30,313]：形 新規の，新しい
- □ **vaccine** [10,085]：名 ワクチン
- □ **develop** [38,325]：動 開発する，発症する，発達する
- □ **become** [12,204]：動 ～になる
- □ **available** [10,780]：形 利用できる，入手できる
- □ **in the future** [430]：将来には

訳 バイオテクノロジー革命ゆえに，多くの新規ワクチンが開発され，そして将来利用可能になるであろう

281 Due to the lack of adequate animal models, the function of these molecules in disease pathogenesis remains poorly understood.

- □ due to～［18,077］：～のせいで，（類 because of～［7,878］，owing to～［782］），～のせいである

  Due to ～［461］：～のせいで，～ゆえに
- □ lack［23,888］：名 欠如 / 動 欠く
- □ adequate［1,554］：形 適切な，適当な（類 appropriate［5,299］，proper［3,218］）
- □ animal model［3,396］：動物モデル
- □ function［92,343］：名 機能 / 動 機能する
- □ molecule［36,845］：名 分子
- □ disease pathogenesis［335］：病因
- □ remain［26,736］：動 ～のままである
- □ poorly understood［3,015］：あまり理解されていない

訳 適切な動物モデルの欠如のせいで，病因におけるこれらの分子の機能はあまり理解されていないままである

282 Imaging technology has advanced the accuracy of diagnosis owing to increased sensitivity.

- □ imaging［11,754］：名 イメージング，造影
- □ technology［4,040］：名 テクノロジー，科学技術
- □ advance［8,632］：動 進行させる（類 progress［4,053］，proceed［2,472］，improve［16,767］）/ 名 進歩
- □ accuracy［3,707］：名 精度，正確さ（類 precision［1,186］）
- □ diagnosis［8,763］：名 診断
- □ owing to ～［782］：～ゆえに，～のせいで（類 because of［7,878］，due to［18,077］）
- □ increased［73,539］：増大した
- □ sensitivity［17,139］：名 感度，感受性

訳 イメージングテクノロジーは，増大した感受性ゆえに診断の精度を進歩させてきた

V つなぎの表現

3 まとめの文例

A. 副詞／副詞的熟語

文例No			用例数
198	Collectively,	まとめると，	1,826
283	Taken together,	まとめると，	5,295
284	In summary,	要約すると，	1,686
285	In conclusion,	まとめると，結論として	2,458

283 **Taken together,** these results demonstrate that alterations of mitochondria potentially contribute to aging and age-related disease in the nervous system.

- □ Taken together, [5,295]：まとめると，（類 in summary [1,701]，in conclusion [2,469]）
- □ **demonstrate that ～** [45,052]：～ということを実証する
- □ **alteration** [9,317]：名 変化（類 change [63,291]，shift [12,913]，conversion [5,249]，modification [9,701]，variation [10,447]）
- □ **mitochondria** [4,876]：名 ミトコンドリア
- □ **potentially** [6,206]：副 潜在的に（類 latently [556]）
- □ **contribute to ～** [19,994]：～に寄与する
- □ **aging** [2,738]：名 加齢，老化
- □ **age-related** [1,477]：年齢関連性の
- □ **disease** [71,437]：名 疾患
- □ **nervous system** [6,242]：神経系

訳 まとめると，これらの結果はミトコンドリアの変化が神経系における加齢および年齢関連疾患に潜在的に寄与することを実証する

284 **In summary,** a number of critical questions remain unanswered concerning the interactions between alcohol and hepatitis C.

- □ In summary, [1,686]：まとめると，（類 taken together [5,918]，in conclusion [2,469]）

□ a number of ～ [7,385]：いくつかの～

□ critical [21,712]：形 決定的に重要な

□ remain [26,736]：動 ～のままである

□ unanswered [167]：形 未解決の（類 unresolved [527], unsolved [80]）

□ concerning [1,266]：前 ～に関する（類 regarding [3,498], with respect to [4,188], in relation to [1,190]）

□ interaction between ～ [11,513]：～の間の相互作用

□ alcohol [5,065]：名 アルコール

□ hepatitis [4,718]：名 肝炎

訳 まとめると，アルコールとC型肝炎の間の相互作用に関して，いくつかの決定的に重要な疑問が未解決のままである

285 **In conclusion,** progenitor cell accumulation in damaged livers has been clearly shown to be modulated by stress-related sympathetic activity.

□ In conclusion, [2,458]：まとめると（類 taken together [5,918], in summary [1,701]），結論として

□ progenitor cell [2,770]：前駆細胞

□ accumulation [13,180]：名 蓄積

□ damaged [1,708]：形 障害された

□ clearly [3,471]：副 明確に，明らかに（類 apparently [3,270], distinctly [473], evidently [199], obviously [110], unambiguously [351]）

□ shown to ～ [17,485]：～することが示される

□ modulate [13,347]：動 調節する

□ stress-related [194]：ストレス関連性の

□ sympathetic [2,448]：形 交感神経の

□ activity [33,998]：名 活動，活性

訳 まとめると，障害された肝臓における前駆細胞蓄積はストレス関連交感神経活動によって調節されることが明確に示されている

4 条件の文例

A. 条件

文例No			用例数
286	If ～	もし～ならば	2,445
287	even if ～	たとえ～だとしても	369
358	unless ～	～でない限りは	816
288	once ～	いったん～すると，一度	2,967
289	Given ～	～を考慮に入れると，～を考えれば	9,905

286 **If** any of these reaction steps are rate-limiting to turnover, the substrate for the rate-limiting reaction should accumulate.

- □ **If** [2,445]：もし～ならば
- □ **reaction** [31,436]：名 反応
- □ **rate-limiting** [2,156]：形 律速の
- □ **turnover** [4,340]：名 代謝回転
- □ **substrate for ～** [2,770]：～に対する基質
- □ **accumulate** [6,498]：動 蓄積する

訳 もしこれらの反応段階のどれかが代謝回転に対して律速であれば，その律速反応に対する基質が蓄積するはずである

287 The physician should not rule out the possibility of an adverse drug reaction in the differential diagnosis **even if** none has been reported previously for the particular drug.

- □ **physician** [4,789]：名 内科医，医師
- □ **rule out** [750]：除外する（類 exclude [3,158]，except [3,308]）
- □ **possibility** [6,450]：名 可能性
- □ **adverse drug reaction** [22]：薬物有害反応
- □ **differential diagnosis** [170]：鑑別診断
- □ **even if ～** [369]：たとえ～だとしても

☐ report [54,972]：動 報告する / 名 報告

☐ previously [34,828]：副 以前に

☐ particular [7,937]：形 特定の

訳 たとえ特定の薬剤について以前に何も報告されていないとしても，内科医は鑑別診断において薬物有害反応の可能性を除外すべきでない

288 **Once** the genetic model has been established, the facts based on it can be clearly explained to the patient and family.

☐ once [2,967]：いったん～すると，一度

☐ genetic [28,358]：形 遺伝的な

☐ establish [19,016]：動 確立する

☐ fact [3,614]：名 事実 （類 finding [35,256]，discovery [3,618]，observation [14,035]）

☐ based on ～ [19,105]：～に基づいた

☐ clearly [3,471]：副 明確に，明らかに

☐ explain [9,556]：動 説明する

☐ family [41,110]：名 家族，ファミリー （類 relative [22,388]）

訳 いったん遺伝的モデルが確立されると，それに基づいた事実が患者と家族に明確に説明されうる

289 **Given** limited advances in renal cell carcinoma, a thorough understanding and testing of rationally targeted agents is needed.

☐ Given [9,905]：～を考慮に入れると，～を考えれば

☐ limited [10,640]：形 限られた

☐ advance [8,632]：名 進歩 / 動 進行させる

☐ renal cell carcinoma [362]：腎細胞癌

☐ thorough [244]：形 徹底的な （類 complete [12,150]，dramatic [3,523]）

☐ understanding [12,133]：名 理解（類 knowledge [4,257]，realization [186]，awareness [541]）

☐ rationally [163]：副 合理的に （類 reasonably [319]）

☐ targeted [8,268]：標的とされた

☐ **agent** [18,880]：名 薬剤，作用物質

☐ **need** [12,039]：動 必要とする / 名 必要性

訳 腎細胞癌における限られた進歩を考慮に入れると，合理的に標的とされた薬剤の徹底的な理解とテストが必要とされる

VI. その他の表現

その他の表現として，ここでは論文でよく使われる**比較の表現**と**asを用いた表現**とを取り上げる．

1 比較の表現の文例

比較の表現としては，thanを用いる比較級の他に，～-fold，%，timesを使うものや，compared，comparison，relativeを使う場合や，to a lesser ～などを用いる表現がある．

A. ～ than

文例No			用例数
290	more than ～	～以上の，より多くの	8,983
291	rather than ～	～よりむしろ	7,606
292	less than ～	～未満の	5,839
293	greater than ～	～以上の，より大きな	5,171
294	higher than ～	～より高い	3,712
295	lower than ～	～より低い	2,760
296	other than ～	～以外の	2,199

290 **More than** 15% of those older than 65 years are reported to be suffering from severe depression.

- ☐ more than ～ [8,983]：～以上の，より多くの
- ☐ older [4,766]：より高齢の
- ☐ reported to ～ [2,484]：～すると報告される
- ☐ suffer from ～ [517]：～を患う
- ☐ severe [12,151]：形 重篤な，重症の，激しい
- ☐ depression [5,911]：名 うつ（病），抑圧

訳 66歳以上の高齢の人々の15％以上は，重篤なうつ病を患っていることが報告されている

291 Age **rather than** menopausal status is a significant independent predictor of vascular events in women with systemic lupus erythematosus.

- □ rather than ～ [7,606]：～よりむしろ
- □ menopausal [203]：形 閉経期の
- □ status [8,697]：名 状態
- □ significant [43,571]：形 重要な，有意な，著しい
- □ independent [25,628]：形 独立した，依存しない，無関係の
- □ predictor [4,048]：名 予知因子
- □ vascular [14,312]：形 血管の
- □ event [24,512]：名 事象，現象
- □ systemic [7,731]：形 全身性の
- □ lupus erythematosus [702]：エリテマトーデス

訳 閉経状態よりむしろ年齢が，全身性エリテマトーデスの女性における血管事象の重要な独立した予知因子である

292 Physical abuse is a leading cause of serious head injury in children **less than** three years of age.

- □ physical [7,312]：形 身体的な
- □ abuse [1,746]：名 虐待，乱用 / 動 乱用する
- □ leading [9,584]：形 主要な
- □ cause [46,697]：名 原因 / 動 引き起こす
- □ serious [1,674]：形 重篤な
- □ head [4,921]：名 頭部
- □ injury [15,721]：名 傷害，損傷
- □ less than ～ [5,839]：～未満の

訳 身体的虐待は，3歳未満の子供における重篤な頭部傷害の主要な原因である

293 These signal transduction pathways are required for **greater than** 90% of the synergistic response.

- □ signal transduction pathway [2,354]：シグナル伝達経路

☐ required for ～ [29,167]：～に必要とされる

☐ greater than ～ [5,171]：～以上の，～より大きな

☐ synergistic [1,887]：形 相乗的な

☐ response [104,667]：名 応答，反応

訳 これらのシグナル伝達経路は，相乗的な応答の90％以上に必要とされる

294 Practicing physicians should be aware that the rate of cardiac complications after smallpox vaccination has been **higher than** expected.

☐ practicing physician [34]：開業医

☐ aware [268]：形 気付いている

☐ rate [59,576]：名 割合，速度

☐ cardiac [17,231]：形 心臓の

☐ complication [5,903]：名 合併症

☐ smallpox [260]：名 天然痘

smallpox vaccination [56]：種痘

☐ higher than ～ [3,712]：～より高い

☐ than expected [564]：予想されるより

訳 開業医は，種痘の後の心臓合併症の割合が予想されるよりも高いことに気付いているべきだ

295 A single intraperitoneal injection of this iron regulatory hormone resulted in iron levels 80% **lower than** in control mice.

☐ single [41,188]：形 一度の，単一の

☐ intraperitoneal [924]：形 腹腔内の

☐ injection [12,342]：名 注射，注入

☐ iron [10,455]：名 鉄

☐ regulatory [19,642]：形 調節性の，制御の

☐ hormone [10,499]：名 ホルモン

☐ result in ～ [48,455]：～という結果になる

☐ lower than ～ [2,760]：～より低い

☐ **control** [78,085]：名 対照群，コントロール，制御 / 動 制御する

訳 この鉄調節性ホルモンの一度の腹腔内注射は，対照群マウスにおけるよりも80％低い鉄レベルという結果になった

296 The aim of this study was to examine the influence of age at menopause on specific causes of death **other than** coronary heart disease.

☐ **aim** [4,574]：名 目的 / 動 目的とする

☐ **examine** [39,969]：動 調べる（類 investigate [30,619]，study [167,914]，test [45,529]）

☐ **influence** [16,903]：名 影響 / 動 影響する

☐ **menopause** [320]：名 閉経，閉経期

☐ **specific** [85,296]：形 特異的な

☐ **cause** [46,697]：名 原因 / 動 引き起こす

☐ **death** [26,658]：名 死

☐ **other than ～** [2,199]：～以外の

☐ **coronary heart disease** [1,129]：冠動脈心疾患

訳 この研究の目的は，冠動脈心疾患以外の死の特異的な原因に対する閉経時の年齢の影響を調べることであった

B. ～-fold, %, times

例文 No			用例数
297	～-fold increase	～倍の増大	1,564
298	～-fold decrease	～倍の低下	278
–	～-fold reduction	～倍の低下	222
299	～% reduction	～%の低下	740
–	～% increase	～%の増大	493
–	～% decrease	～%の低下	438
300	increased ～-fold	～倍増大した	265
–	increased ～%	～%増大した	124
301	～-fold higher	～倍高い	987
–	～-fold more …	～倍より…	614
302	～-fold more potent	～倍より強力な	103
–	～-fold lower	～倍より低い	478
–	～-fold greater	～倍より大きい	448
295	～% lower	～%より低い	284
–	～% higher	～%より高い	243
–	～ times more …	～倍より…	899
303	～ times more likely to …	～倍より…しそうな	136
304	～ times higher	～倍より高い	521

297 This novel fluorescence assay provided a **20-fold increase** in sensitivity with a broad dynamic range.

- ☐ **novel** [30,313]：形 新規の，新しい
- ☐ **fluorescence** [11,995]：名 蛍光
- ☐ **assay** [38,173]：名 アッセイ，分析 / 動 アッセイする
- ☐ **provide** [56,723]：動 提供する，与える
- ☐ **～-fold increase** [3,518]：～倍の増大
- ☐ **sensitivity** [17,139]：名 感度，感受性

☐ **broad** [4,339]：形 広い，広範な（類 wide [7,526]，widespread [2,588]）

☐ **dynamic** [5,935]：形 ダイナミックな，動的な

☐ **range** [25,304]：名 レンジ，範囲 / 動 わたる

訳 この新規の蛍光アッセイは広いダイナミックレンジを持ち，感度の20倍の増大をもたらした

298 A subset of these genes exhibited an approximately **2-fold decrease** in retinal gene expression.

☐ **subset** [8,910]：名 サブセット

☐ **exhibit** [27,339]：動 示す

☐ **approximately** [34,275]：副 およそ（類 roughly [762]，about [19,601]，ca [20,024]）

☐ **～-fold decrease** [644]：～倍の低下

☐ **retinal** [8,244]：形 網膜の

☐ **gene expression** [21,171]：遺伝子発現

訳 これらの遺伝子のサブセットは，網膜の遺伝子発現のおよそ2倍の低下を示した

299 A **45% reduction** in accumulation of these cells was observed in an acute lung injury model.

☐ **～% reduction** [1,480]：～%の低下

☐ **accumulation** [13,180]：名 蓄積

☐ **observe** [51,935]：動 観察する

☐ **acute lung injury** [452]：急性肺傷害

訳 これらの細胞の蓄積の45%の低下が急性肺傷害モデルにおいて観察された

300 Although the men and women showed similar insulin responses to feeding, leptin concentrations **increased 3-fold** only in women.

☐ **show** [148,875]：動 示す

☐ **similar** [45,428]：形 類似の，似ている

☐ insulin [20,353]：名 インスリン

☐ response [104,667]：名 応答，反応

☐ feeding [2,867]：名 摂食

☐ leptin [3,481]：名 レプチン

☐ concentration [39,910]：名 濃度，集中

☐ increased ～-fold [550]：～倍増大した

訳 男性と女性は摂食に対する類似のインスリン応答性を示したが，レプチン濃度は女性においてのみ３倍増大した

301 While low doses dramatically enhanced immune responses, **10-fold higher** doses did not augment responses.

☐ low [39,773]：形 低い（反 high [68,406]）

☐ dose [29,485]：名 用量，投与量

☐ dramatically [4,756]：副 劇的に（類 markedly [7,503], significantly [48,939], drastically [559], strongly [11,204]）

☐ enhanced [20,161]：増強した

☐ immune response [8,912]：免疫応答

☐ ～-fold higher [2,030]：～倍高い

☐ augment [3,582]：動 増大させる，増強する

☐ response [104,667]：名 応答，反応

訳 低い用量は免疫応答を劇的に増強したけれども，１０倍高い用量は応答を増大させなかった

302 Although azide and formate exhibited different potencies as scavengers, the former was almost **2-fold more potent** than the latter.

☐ azide [519]：名 アジ化物

☐ formate [303]：名 ギ酸

☐ exhibit [27,339]：動 示す

☐ different [48,428]：形 異なる（類 distinct [22,633], dissimilar [331], disparate [647]）

☐ potency [3,727]：名 効力

☐ **scavenger** [1,003]：名 スカベンジャー

☐ **former** [1,250]：名 前者 / 形 前の

☐ **almost** [4,947]：副 ほとんど

☐ **～-fold more potent** [218]：～倍より強力な

☐ **latter** [4,473]：名 後者 / 形 後者の

訳 アジ化物とギ酸はスカベンジャーとして異なる効力を示したが，前者は後者よりほとんど2倍ほどより強力だった

303 Females are two to three **times more likely to** be affected by rheumatoid arthritis than males.

☐ **female** [9,873]：名 女性，雌

☐ **～ times more…** [899]：～倍より…

～ times more likely to … [136]：～倍より…しそうな

☐ **affected** [11,803]：罹患した

☐ **rheumatoid arthritis** [1,150]：関節リウマチ

☐ **male** [12,104]：名 男性，雄（類 man [10,301]）

訳 女性は男性より関節リウマチに2～3倍，より罹患しそうである

304 The AIDS rates for Hispanic men were reported to be three **times higher** than for white men in the US.

☐ **AIDS** [2,994]：名 エイズ

☐ **rate** [59,576]：名 割合，速度

☐ **Hispanic** [888]：形 ヒスパニックの，ラテンアメリカ系の

☐ **reported to ～** [2,484]：～すると報告される

☐ **～ times higher** [521]：～倍より高い

☐ **white** [5,336]：形 白い，白人の

訳 合衆国においてヒスパニックの男性のエイズの割合は，白人男性よりも3倍高いと報告された

C. compared, comparison, relative

文例No			用例数
305	Compared with ～	～と比べて	2,369
–	Compared to ～	～と比べて	636
306	in comparison with ～	～と比べて	939
–	in comparison to ～	～と比べて	409
307	relative to ～	～と比べて	35,718

305 **Compared with** wild-type mice, these mice displayed markedly exacerbated disease progression and pathology.

- □ **Compared with ～** [2,369]：～と比べて
- □ **wild-type** [39,564]：野生型の
- □ **display** [15,251]：動 示す，提示する / 名 提示
- □ **markedly** [7,503]：副 顕著に，著しく
- □ **exacerbate** [943]：動 悪化させる
- □ **disease** [71,437]：名 疾患
- □ **progression** [12,698]：名 進行
- □ **pathology** [3,222]：名 病態，病理，病理学

訳 野生型マウスと比べて，これらのマウスは顕著に悪化した疾患の進行と病態を示した

306 Screening tests for lung cancer with annual chest x-ray films and sputum cytology resulted in no improvement in overall mortality **in comparison with** control subjects.

- □ **screening** [3,604]：名 スクリーニング
- □ **lung cancer** [2,400]：肺癌
- □ **annual** [1,178]：形 年一回の，年間の
- □ **chest** [1,668]：名 胸部
- □ **film** [2,421]：名 フィルム
- □ **sputum** [498]：名 痰
- □ **cytology** [207]：名 細胞診

☐ **result in ～** [48,455]：～という結果になる

☐ **improvement** [6,188]：名 改善

☐ **overall** [14,048]：形 全体の

☐ **mortality** [12,864]：名 死亡率

☐ **in comparison with ～** [939]：～と比べて

☐ **control** [78,085]：名 対照群，コントロール，制御 / 動 制御する

☐ **subject** [26,037]：名 被験者，対象 / 動 受けさせる

訳 年一回の肺X線フィルムおよび喀痰細胞診による肺癌のスクリーニングテストは，対照群被検者と比べて全体の死亡率の改善という結果にはならなかった

307 Rodent models of obesity showed significant increases in food intake and weight gain **relative to** controls.

☐ **rodent** [2,818]：名 げっ歯類

rodent model [293]：げっ歯類モデル

☐ **obesity** [3,240]：名 肥満

☐ **show** [148,875]：動 示す

☐ **significant** [43,571]：形 有意な，著しい，重要な

☐ **increase in ～** [35,718]：～の増大

☐ **food intake** [933]：食物摂取

☐ **weight gain** [753]：体重増加

☐ **relative to ～** [8,289]：～と比べて

☐ **control** [78,085]：名 対照群，コントロール，制御 / 動 制御する

訳 肥満のげっ歯類モデルは，対照群と比べて食餌摂取と体重増加の有意な増大を示した

D. to a lesser ～／… degree of ～

文例No			用例数
308	to a lesser extent	より少ない程度で	1,069
－	to a greater extent	より大きな程度で	337
309	a high degree of ～	高度の～	867
137	to a lesser degree	より少ない程度で	240

308 Histologically, iron is found in increased amounts in organs such as the liver, heart, and pancreas and **to a lesser extent** in the endocrine glands.

- □ **histologically** [941]：副 組織学的に
- □ **iron** [10,455]：名 鉄
- □ **increase** [146,670]：動 増大させる，増大する，名 増大
- □ **amount** [11,659]：名 量
- □ **organ** [10,264]：名 臓器，器官
- □ **heart** [20,807]：名 心臓
- □ **pancreas** [2,322]：名 膵臓
- □ **to a lesser extent** [1,069]：より少ない程度で
- □ **endocrine** [1,135]：名 内分泌
- □ **gland** [4,438]：名 腺

訳 組織学的に，鉄は肝臓，心臓，膵臓のような臓器およびより少ない程度で内分泌腺において量の増大が見つけられる

309 Medications with a low therapeutic index known to have a high risk of drug interactions should raise **a high degree of** suspicion of adverse drug interactions.

- □ **medication** [2,850]：名 薬物療法
- □ **therapeutic index** [128]：治療係数
- □ **known to ～** [8,375]：～すると知られている
- □ **risk** [37,608]：名 リスク，危険
- □ **drug interaction** [154]：薬物相互作用

- □ **raise** [5,926]：動 提起する，上げる
- □ **a high degree of ～** [919]：高度の～
- □ **suspicion** [141]：名 疑い
- □ **adverse** [4,131]：形 有害な

訳 薬物相互作用の高いリスクを持つことが知られている低い治療係数の薬物療法は，有害な薬物相互作用に高度の疑念を提起するはずである

2 asを用いた表現の文例

asは意味も用法も非常に多様である．時や理由を表す節を導くだけでなく，「～として」という意味や，as ～ asの表現などがよく用いられる．

A. as＋過去分詞

文例No			用例数
310	as compared with ～	～と比較すると	2,998
311	as determined by ～	～によって決定されるように	2,347
312	as measured by ～	～によって測定されるように	2,230
313	as expected	予想されるように	1,235
220	as assessed by ～	～によって評価されるように	1,146
314	as evidenced by ～	～によって証明されるように	1,038
315	as shown by ～	～によって示されるように	999
316	as indicated by ～	～によって示されるように	926
317	as demonstrated by ～	～によって実証されるように	888
318	as judged by ～	～によって判断されるように	627
319	as opposed to ～	～とは対照的に	611
320	as observed	観察されるように	500
321	as detected by ～	～によって検出されるように	426
322	as revealed by ～	～によって明らかにされるように	403
323	as defined by ～	～によって定義される（ように）	358
324	as described	述べられているように	219
325	as predicted by ～	～によって予想されるように	211
326	as reflected by ～	～によって反映されるように	198

310 **As compared with** placebo, both classes of drugs demonstrated efficacy in the treatment of rheumatoid arthritis.

□ **as compared with ～** [2,998]：～と比較すると
□ **placebo** [7,595]：名 プラセボ，偽薬
□ **demonstrate** [80,078]：動 実証する
□ **efficacy** [8,112]：名 有効性
□ **treatment** [60,138]：名 治療，処理
□ **rheumatoid arthritis** [1,150]：関節リウマチ

訳 偽薬と比較すると，両方のクラスの薬物は関節リウマチの治療における有効性を実証した

311 DEC1 physically interacted with Bmal1, **as determined by** coimmunoprecipitation.

□ **physically** [1,705]：副 物理的に
□ **interact with ～** [14,220]：～と相互作用する
□ **as determined by ～** [2,347]：～によって決定されるように
□ **coimmunoprecipitation** [839]：名 免疫共沈降

訳 免疫共沈降によって決定されたように，DEC1はBmal1と物理的に相互作用した

312 These osteoblasts were shown to produce nerve growth factor, **as measured by** ELISA.

□ **osteoblast** [1,608]：名 骨芽細胞
□ **shown to ～** [17,485]：～することが示される
□ **produce** [38,705]：動 産生する（類 generate [28,107]，yield [12,522]），引き起こす
□ **nerve growth factor** [819]：神経成長因子
□ **as measured by ～** [2,230]：～によって測定されるように

☐ ELISA [2,092]：名 ELISA 法，酵素結合免疫吸着検定法

訳 ELISA 法によって測定されたように，これらの骨芽細胞は神経成長因子を産生することが示された

313 As expected, these young mice showed markedly elevated apoptosis compared with wild-type controls.

☐ as expected [1,235]：予想されるように

☐ young [4,725]：形 若い

☐ show [148,875]：動 示す

☐ markedly [7,503]：副 顕著に，著しく（類 remarkably [2,458]，strikingly [1,715]，dramatically [4,756]，notably [1,715]，extremely [2,414]，significantly [48,939]）

☐ elevate [13,039]：動 上昇させる（類 up-regulate [3,805]，augment [3,582]，increase [146,670]）

☐ apoptosis [31,507]：名 アポトーシス

☐ compared with ～ [32,460]：～と比較して

☐ wild-type [39,564]：野生型の

☐ control [78,085]：名 対照群，コントロール，制御 / 動 制御する

訳 予測されたように，これらの若いマウスは野生型の対照群と比較して顕著に上昇したアポトーシスを示した

314 Environmental factors are considered important in the etiology of gastric cancer, as evidenced by Japanese descendents in Hawaii showing a decreased incidence of gastric cancer.

☐ environmental factor [629]：環境因子

☐ consider [8,440]：動 考える（類 regard [2,158]，think [8,004]，believe [3,698]），考慮する（類 take into account [598]）

☐ important [45,928]：形 重要な

☐ etiology [1,851]：名 病因，病因論

☐ gastric [3,253]：形 胃の

☐ as evidenced by ～ [1,038]：～によって証明されるように

☐ descendant [205]：名 子孫（類 descendent [35]，offspring [1,705]，progeny [1,604]）

VI その他の表現

- ☐ **show** [148,875]：動 示す
- ☐ **decreased** [26,518]：低下した
- ☐ **incidence** [8,627]：名 発生率，頻度

訳 ハワイの日本人の子孫が低下した胃癌の発生率を示していることによって証明されるように，環境因子は胃癌の病因において重要であると考えられる

315 These symptoms are accompanied by increasing evidence of damage in the basal ganglia **as shown by** magnetic resonance imaging.

- ☐ **symptom** [9,302]：名 症状
- ☐ **accompanied by ～** [4,941]：～に伴われる，～を伴う
- ☐ **increasing** [11,608]：増大する
- ☐ **evidence** [35,499]：名 証拠／動 立証する
- ☐ **damage** [15,696]：名 障害，損傷 / 動 損傷する
- ☐ **basal ganglia** [840]：基底核
- ☐ **as shown by ～** [999]：～によって示されるように
- ☐ **magnetic resonance imaging** [1,609]：磁気共鳴画像法（MRI）

訳 磁気共鳴画像法によって示されるように，これらの症状は基底核における増大する障害の証拠を伴う

316 One of the symptoms of dementia is impairment in abstract thinking, **as indicated by** difficulty in defining words and concepts previously known.

- ☐ **symptom** [9,302]：名 症状
- ☐ **dementia** [1,681]：名 認知症
- ☐ **impairment** [3,806]：名 障害（類 dysfunction [6,236]，disturbance [970]，damage [13,052]，disorder [17,713]）
- ☐ **abstract thinking** [1]：抽象思考
- ☐ **as indicated by ～** [926]：～によって示されるように
- ☐ **difficulty** [1,608]：名 困難（性）
- ☐ **define** [22,878]：動 定義する，明らかにする，規定する
- ☐ **word** [953]：名 言葉

☐ concept [2,660]：名 概念

☐ previously [34,828]：副 以前に

訳 以前に知られていた言葉や概念を定義する際の困難性によって示されるように，認知症の症状のひとつは抽象思考の障害である

317 This effect is gene-specific, **as demonstrated by** the fact that G3PDH expression is not affected.

☐ effect [106,593]：名 効果，影響

☐ gene-specific [364]：遺伝子特異的な

☐ **as demonstrated by ～** [888]：～によって実証されるように

☐ the fact that ～ [2,097]：～という事実

☐ expression [154,475]：発現

☐ affect [32,553]：動 影響する

訳 G3PDH発現は影響されないという事実によって実証されるように，この効果は遺伝子特異的である

318 These three binding sequences accounted for about 65% of the transcriptional effect, **as judged by** transient transfection assays.

☐ binding [133,851]：名 結合

☐ sequence [91,510]：名 配列 / 動 配列決定する

☐ account for ～ [8,259]：～を占める，～を説明する

☐ transcriptional [21,243]：形 転写の

☐ effect [106,593]：名 効果，影響

☐ **as judged by ～** [627]：～によって判断されるように

☐ transient [9,961]：形 一過性の

☐ transfection [5,761]：名 形質移入，トランスフェクション

☐ assay [38,173]：名 アッセイ，分析 / 動 アッセイする

訳 一過性の形質移入アッセイによって判断されたように，これらの3つの結合配列は転写効果のおよそ65％を占めた

319 Here, we report studies detailing the cross-sectional imaging of bowel obstruction **as opposed to** the more traditional imaging with barium radiography.

- □ **report** [54,972]：動 報告する / 名 報告
- □ **detail** [7,214]：動 詳しく調べる / 名 詳細
- □ **cross-sectional** [1,332]：横断面の
- □ **imaging** [11,754]：名 造影，イメージング
- □ **bowel** [1,636]：名 腸（類 intestine [2,328]，gut [1,699]）
- □ **obstruction** [1,089]：名 閉塞（類 occlusion [2,391]，blockage [377]）
- □ **as opposed to ～** [611]：～とは対照的に
- □ **traditional** [2,102]：形 伝統的な
- □ **barium** [273]：名 バリウム
- □ **radiography** [340]：名 X 線検査

訳 ここに我々は，バリウムX線検査によるより伝統的な造影法とは対照的に，腸閉塞の横断造影を詳しく調べる研究を報告する

320 Besides ultraviolet radiation from sun exposure, there appear to be other causal agents in the development of cutaneous melanoma **as observed** in epithelial carcinomas.

- □ **besides** [496]：前 ～の他に，～に加えて
- □ **ultraviolet** [1,656]：名 紫外線
- □ **radiation** [7,553]：名 照射（類 irradiation [3,658]）
- □ **sun** [389]：名 太陽
- □ **exposure** [16,765]：名 曝露
- □ **appear to ～** [21,779]：～するように思われる
- □ **causal** [915]：形 原因となる
- □ **agent** [18,880]：名 薬剤，作用物質，因子
- □ **development** [48,509]：名 発症，発生，開発
- □ **cutaneous** [1,826]：形 皮膚の（類 dermal [1,113]）
- □ **melanoma** [5,108]：名 メラノーマ，黒色腫
- □ **as observed** [500]：観察されるように

☐ epithelial [16,204]：形 上皮の

☐ carcinoma [9,238]：名 癌腫（類 cancer [45,184]）

訳 上皮癌腫において観察されるように，太陽曝露からの紫外線照射の他に皮膚のメラノーマの発症における他の原因因子があるように思われる

321 Unilateral removal of the avian cochlea caused a drastic reduction in the expression of these proteins in the NM neurons **as detected by** immunocytochemistry.

☐ unilateral [1,135]：形 一側性の，片側の

☐ removal [5,555]：名 除去（類 elimination [2,643]，abstraction [466]）

☐ avian [1,561]：形 トリの

☐ cochlea [420]：名 蝸牛

☐ cause [46,697]：動 引き起こす / 名 原因

☐ drastic [268]：形 強烈な（類 strong [11,381]）

☐ reduction in ～ [12,245]：～の低下

☐ expression [154,475]：名 発現

☐ neuron [39,544]：名 ニューロン

☐ as detected by ～ [426]：～によって検出されるように

☐ immunocytochemistry [1,231]：名 免疫細胞化学

訳 免疫細胞化学によって検出されたように，トリの蝸牛の一側性の除去はNMニューロンにおけるこれらのタンパク質の発現の強烈な低下を引き起こした

322 Mutants in this particular gene exhibited asynchronous overreplication during normal growth, **as revealed by** flow cytometry.

☐ mutant [79,727]：名 変異体 / 形 変異の

☐ particular [7,937]：形 特定の

☐ exhibit [27,339]：動 示す

☐ asynchronous [277]：形 非同期性の

☐ overreplication [19]：名 過剰複製

☐ normal [44,556]：形 正常な

☐ growth [59,775]：名 成長，増殖

☐ as revealed by ～ [403]：～によって明らかにされるように

☐ flow cytometry [1,901]：フローサイトメトリー

訳 フローサイトメトリーによって明らかにされたように，この特定の遺伝子の変異は正常な成長の間の非同期性の過剰複製を示した

323 The mean age of onset of puberty for girls **as defined by** breast budding is a little earlier than that for boys **as defined by** enlargement of testes.

☐ mean age [1,587]：平均年齢

☐ onset [9,566]：名 開始，発症

☐ puberty [227]：名 思春期

☐ as defined by ～ [358]：～によって定義される（ように）

☐ breast [14,198]：名 乳房，胸

☐ budding [2,056]：名 発芽

☐ earlier [4,774]：形 より早い（early の比較級）

☐ enlargement [676]：名 拡大（類 expansion [6,180]，extension [4,585]）

☐ testis [2,031]：名（複 testes）精巣

訳 乳房の発芽で定義される少女の思春期の開始の平均年齢は，精巣の拡大で定義される少年のそれよりも少し早期である

324 **As described** in the accompanying article, our study represents the first evidence of a super-integron in a non-pathogenic bacterium.

☐ as described [219]：述べられているように

☐ accompanying [843]：付随する

☐ article [3,606]：名 論文（類 paper [4,717]），記事

☐ represent [17,123]：動 示す，表す

☐ evidence [35,499]：名 証拠 / 動 立証する

☐ integron [49]：名 インテグロン

☐ pathogenic [3,726]：形 病原性のある（類 virulent [1,322]）

☐ bacterium [11,218]：名（複 bacteria）細菌

訳 付随論文で述べられているように，我々の研究は非病原性バクテリアにおけるスーパー・インテグロンの最初の証拠を示す

325 **As predicted by** the model, we found genetic evidence that these cells are sufficient to alter neural activity in regions involved in autonomic and neuroendocrine control.

- □ as predicted by ～ [211]：～によって予想されるように
- □ genetic [28,358]：形 遺伝的な
- □ evidence that ～ [9,965]：～という証拠
- □ sufficient to ～ [5,852]：～するのに十分な
- □ alter [21,411]：動 変化させる
- □ neural [11,068]：形 神経の（類 nervous [6,387]）
- □ activity [33,998]：名 活動，活性
- □ region [77,257]：名 領域
- □ involved in ～ [25,574]：～に関与する
- □ autonomic [1,073]：形 自律神経性の
- □ neuroendocrine [937]：名 神経内分泌
- □ control [78,085]：名 制御，コントロール，対照群 / 動 制御する

訳 そのモデルによって予想されるように，我々は，これらの細胞が自律神経性および神経内分泌性制御に関与する領域における神経活動を変えるのに十分であるという遺伝的な証拠を見つけた

326 These results underscore the role of this pathway in the biology of plasma cell growth **as reflected by** its influence on survival.

- □ underscore [1,128]：動 強調する（類 emphasize [1,584]，highlight [2,970]）
- □ role [87,150]：名 役割
- □ pathway [59,666]：名 経路
- □ biology [3,565]：名 生物学
- □ plasma cell [770]：プラズマ細胞
- □ growth [59,775]：名 増殖，成長
- □ as reflected by ～ [198]：～に反映されるように

VI その他の表現

□ **influence** [16,903]：名 影響（類 effect [107,038]，impact [6,642]）/ 動 影響する

□ **survival** [28,139]：名 生存，生存率

訳 これらの結果は，生存に対するそれの影響に反映されるように，プラズマ細胞増殖の生物学におけるこの経路の役割を強調する

B. as ＋名詞

文例No			用例数
327	as a result of ～	～の結果として	2,540
328	as part of ～	～の一部として	1,670
329	as a consequence of ～	～の結果として	855
330	as a substrate	基質として	583
331	as a means	手段として	481
332	as a marker	マーカーとして	440
333	as controls	対照群として，コントロールとして	405
334	as a target	標的として	368
335	as a probe	プローブとして	353
336	as a whole	全体として	348
337	as a mechanism	機構として	339
338	as a source of ～	～の原因として，～の源として	311
339	as an inhibitor	阻害剤として	295
340	as a component of ～	～の構成成分として	294
341	as a first step	最初の段階として	247
342	as an example	例として	207
343	as an index of ～	～の指標として	189
344	as an alternative to ～	～の代替として	159
345	as the basis for ～	～の基礎として	141

327 Family members taking care of a dementia patient sometimes develop clinically significant symptoms of depression **as a result of** the stress associated with prolonged caregiving.

- □ **take care of ～** [7]：～を世話する
- □ **dementia** [1,681]：名 認知症
- □ **sometimes** [895]：副 ときどき
- □ **develop** [38,325]：動 発症する，開発する，発達する
- □ **clinically** [4,700]：副 臨床的に
- □ **significant** [43,571]：形 著しい，重要な，有意な
- □ **symptom** [9,302]：名 症状
- □ **depression** [5,911]：名 うつ，抑圧
- □ **as a result of ～** [2,540]：～の結果として
- □ **stress** [18,967]：名 ストレス
- □ **associated with ～** [54,439]：～と関連する
- □ **prolonged** [5,461]：長引く
- □ **caregiving** [42]：名 介護

訳 認知症の患者を世話する家族のメンバーは，長引く介護に関連したストレスの結果として，臨床的に著しいうつの症状をときどき発症する

328 Many African countries are urged to complete measles supplemental immunization activities in children aged 9 months to 14 years **as part of** a comprehensive measles-control strategy.

- □ **country** [1,519]：名 国
- □ **urge** [89]：動 勧める，せきたてる
- □ **complete** [14,902]：動 完了する / 形 完全な
- □ **measles** [804]：名 麻疹
- □ **supplemental** [401]：形 追加の
- □ **immunization** [4,044]：名 免疫化
- □ **activity** [33,998]：名 活動，活性
- □ **as part of ～** [1,655]：～の一部として

□ comprehensive [1,968]：形 包括的な，網羅的な

□ strategy [13,405]：名 戦略

訳 多くのアフリカの国々は，包括的な麻疹制御戦略の一部として，９カ月から14歳の子供における麻疹の追加の免疫化活動を完了することを勧められている

329 Thrombosis of the renal arteries and segmental branches may arise **as a consequence of** intrinsic pathology of the renal arteries.

□ thrombosis [1,928]：名 血栓症

□ renal artery [321]：腎動脈

□ segmental [967]：形 分節の

□ branch [5,705]：名 枝／動 枝分かれする

□ arise [6,458]：動 生じる，起こる

□ as a consequence of ～ [855]：～の結果として

□ intrinsic [5,126]：形 内因性の

□ pathology [3,222]：名 病態，病理，病理学

訳 腎動脈および分節の枝の血栓症は腎動脈の内因性の病態の結果として生じるかもしれない

330 This long single-stranded tail is expected to serve **as a substrate** for telomerase.

□ long [24,362]：形 長い

□ single-stranded [2,946]：１本鎖の

□ tail [6,756]：名 尾部

□ expected to ～ [1,448]：～すると予想される，～すると期待される

□ serve as ～ [7,135]：～として働く，～として役立つ

□ as a substrate [583]：基質として

□ telomerase [3,604]：名 テロメラーゼ

訳 この長い１本鎖の尾部は，テロメラーゼに対する基質として働くことが予想される

331 A wealth of data for exercise tolerance testing can be used **as a means** of identifying asymptomatic patients at high risk for coronary heart disease.

- □ **wealth of 〜** [224]：大量の〜
- □ **data for 〜** [1,486]：〜のデータ
- □ **exercise tolerance testing** [7]：運動負荷試験
- □ **used as 〜** [4,361]：〜として使われる
- □ **as a means** [481]：手段として
- □ **identify** [73,456]：動 同定する
- □ **asymptomatic** [1,554]：形 無症候性の
- □ **at high risk for 〜** [369]：〜のリスクの高い
- □ **coronary heart disease** [1,129]：冠動脈心疾患

訳 運動負荷試験の大量のデータは，冠動脈心疾患のリスクの高い無症候性の患者を同定する手段として使われうる

332 This antigen is known to serve **as a marker** of HBV infection.

- □ **antigen** [23,859]：名 抗原
- □ **known to 〜** [8,375]：〜すると知られている
- □ **serve as 〜** [7,135]：〜として役立つ，〜として働く
- □ **as a marker** [440]：マーカーとして
- □ **HBV** [1,969]：名 B 型肝炎ウイルス（hepatitis B virus）
- □ **infection** [44,601]：名 感染

訳 この抗原は，B 型肝炎ウイルス感染のマーカーとして役立つことが知られている

333 Sixty-seven subjects without prominent visual symptoms were selected **as controls**.

- □ **subject** [26,037]：名 対象者 / 動 受けさせる
- □ **prominent** [2,934]：形 顕著な
- □ **visual** [8,105]：形 視覚の

☐ symptom [9,302]：名 症状

☐ select [10,123]：動 選択する

☐ as controls [405]：対照群として，コントロールとして

訳 顕著な視覚的な症状のない67名の被験者が対照群として選択された

334 The kidney is a major excretory organ and serves **as a target** for many hormones.

☐ kidney [10,654]：名 腎臓

☐ major [30,348]：形 主要な

☐ excretory [123]：形 排泄の

☐ organ [10,264]：名 器官，臓器

☐ serve as ～ [7,135]：～として働く，～として役立つ

☐ as a target [368]：標的として

☐ hormone [10,499]：名 ホルモン

訳 腎臓は，主要な排泄器官であり，そして多くのホルモンの標的として働く

335 Use of the cDNA **as a probe** enabled us to determine the onset, relative levels, and locations of gene expression in various adult tissues.

☐ use of ～ [17,491]：～の使用

☐ as a probe [353]：プローブとして

☐ enable [5,078]：動 可能にする

☐ determine [58,574]：動 決定する

☐ onset [9,566]：名 開始，発症

☐ relative [21,076]：形 相対的な

☐ location [8,396]：名 部位（類 site [99,321]，area [5,037]，locus [18,724]，region [77,257]）

☐ gene expression [21,171]：遺伝子発現

☐ various [14,339]：形 さまざまな

☐ tissue [45,858]：名 組織

訳 プローブとしてのcDNAの使用は，我々がさまざまな成体組織における遺伝子発現の開始，相対レベルおよび部位を決定することを可能にした

VI その他の表現

336 The method introduced here is practical and available to the target population **as a whole**.

- □ **method** [32,975]：名 方法
- □ **introduce** [6,289]：動 紹介する，導入する
- □ **practical** [1,234]：形 実用的な
- □ **available** [10,780]：形 利用できる，入手できる
- □ **target** [51,095]：名 標的，動 標的にする
- □ **population** [28,214]：名 集団，人口
- □ **as a whole** [348]：全体として

訳 ここで紹介された方法は実用的でそして，全体として標的集団に利用できる

337 We reviewed the evidence of apoptosis **as a mechanism** for this tumor response in p53 mutant breast cancer.

- □ **review** [12,123]：動 概説する / 名 総説，概説
- □ **evidence** [35,499]：名 証拠 / 動 立証する
- □ **apoptosis** [31,507]：名 アポトーシス
- □ **as a mechanism** [339]：機構として
- □ **tumor** [61,098]：名 腫瘍
- □ **response** [104,667]：名 反応，応答
- □ **mutant** [79,727]：形 変異の / 名 変異体
- □ **breast cancer** [8,621]：乳癌

訳 我々は，p53変異乳癌においてこの腫瘍応答の機構として，アポトーシスの証拠を概説した

338 The role of genetic factors **as a source of** interindividual variation in drug response seems to be well established.

- □ **role** [87,150]：名 役割
- □ **genetic factor** [573]：遺伝因子
- □ **as a source of ～** [311]：～の原因として，～の源として
- □ **interindividual** [159]：形 個体間の

☐ variation [10,447]：名 変動

☐ drug response [105]：薬物応答

☐ seem to ～ [3,153]：～するように思われる

☐ establish [19,016]：動 確立する

訳 薬物応答における個体間の変動の原因としての遺伝的因子の役割は，よく確立されているように思われる

339 This gene has been reported to act **as an inhibitor** of neuronal cell proliferation in the secondary neural tube.

☐ reported to ～ [2,484]：～すると報告される

☐ act as ～ [6,441]：～として作用する

☐ as an inhibitor [295]：阻害剤として

☐ neuronal cell [1,319]：神経細胞

☐ proliferation [19,120]：名 増殖

☐ secondary [11,033]：形 二次性の

☐ neural tube [1,040]：神経管

訳 この遺伝子は，二次性神経管における神経細胞増殖の阻害剤として働くことが報告されている

340 This homolog of the protein has been implicated **as a component of** the viral budding machinery.

☐ homolog [6,055]：名 ホモログ

☐ implicate [13,847]：動 関連づける

☐ component [28,501]：名 構成成分

as a component of ～ [294]：～の構成成分として

☐ viral [21,817]：形 ウイルスの，ウイルス性の

☐ budding [2,056]：名 発芽

☐ machinery [3,238]：名 機構（類 mechanism [70,869]，mode [6,316]）

訳 そのタンパク質のこのホモログは，ウイルス発芽機構の構成成分として関連づけられている

341 **As a first step**, we have embarked on the characterization of this enzyme in zebrafish.

- □ as a first step [247]：最初の段階として
- □ embark [30]：動 取り組む（類 address [6,396]）
- □ characterization [6,815]：名 特徴づけ
- □ enzyme [49,164]：名 酵素
- □ zebrafish [2,229]：名 ゼブラフィッシュ

訳 最初の段階として，我々はゼブラフィッシュにおけるこの酵素の特徴づけに取り組んできた

342 **As an example**, we show how contemporary parenteral products with microbial contaminants can be considered safe under current pharmacopoeia tests, but provoke adverse clinical effects.

- □ as an example [207]：例として
- □ show [148,875]：動 示す
- □ contemporary [430]：形 現代の
- □ parenteral [430]：形 非経口的の
- □ product [24,771]：名 産物
- □ microbial [2,207]：形 微生物の
- □ contaminant [363]：名 汚染物質
- □ consider [8,440]：動 考える，みなす
- □ safe [1,678]：形 安全な
- □ current [18,486]：形 現在の
- □ pharmacopoeia [4]：名 薬物類，薬局方
- □ provoke [549]：動 誘発する（類 elicit [7,731]，evoke [5,622]，induce [116,671]）
- □ adverse [4,131]：形 有害な
- □ clinical [31,256]：形 臨床の
- □ effect [106,593]：名 作用，影響，効果

訳 例として，我々は，どのように微生物汚染物質を持つ現代の非経口産物が現在の薬物類テスト下で安全であると考えられながら，有害な臨床作用を引き起こすかを示す

343 These abnormalities found in patients with renal failure have been used **as an index of** the adequacy of hemodialysis.

- ☐ **abnormality** [7,645]：名 異常
- ☐ **patient with ～** [43,273]：～の患者
- ☐ **renal failure** [882]：腎不全
- ☐ **used as ～** [4,361]：～として使われる
- ☐ **as an index of ～** [189]：～の指標として
- ☐ **adequacy** [217]：名 妥当性（類 validity [1,005]，appropriateness [137]）
- ☐ **hemodialysis** [602]：名 血液透析

訳 腎不全の患者にみられるこれらの異常は，血液透析の妥当性の指標として使われてきた

344 Another potent inhibitor of platelet function is being used **as an alternative to** aspirin in patients with cerebrovascular disease.

- ☐ **potent** [11,330]：形 強力な
- ☐ **inhibitor** [43,673]：名 阻害剤
- ☐ **platelet** [10,686]：名 血小板
- ☐ **function** [92,343]：名 機能，動 機能する
- ☐ **used as ～** [4,361]：～として使われる
- ☐ **as an alternative to ～** [159]：～の代替として
- ☐ **aspirin** [1,568]：名 アスピリン
- ☐ **patient with ～** [43,273]：～の患者
- ☐ **cerebrovascular** [485]：形 脳血管の
- ☐ **disease** [71,437]：名 疾患

訳 血小板機能のもうひとつの強力な阻害剤は，脳血管疾患を持つ患者におけるアスピリンの代替として使われている

345 Evidence-based medicine requires scientific methods **as the basis for** understanding and treating disease.

- □ **evidence-based medicine** [33]：根拠に基づく医学
- □ **require** [65,968]：動 必要とする（類 need [12,039]，necessitate [418]，demand [1,297]）
- □ **scientific** [1,056]：形 科学的な
- □ **method** [32,975]：名 方法
- □ **as the basis for ～** [141]：～の基礎として
- □ **understand** [12,582]：動 理解する
- □ **treat** [29,577]：動 治療する，処理する
- □ **disease** [71,437]：名 疾患

訳 根拠に基づく医学は，疾患を理解しそして治療するための基礎として科学的な方法を必要とする

C. as ～ as

文例No			用例数
346	as well as ～	～と同様に	26,236
347	as early as ～	早くも～ (～と同じぐらい早い)	842
348	as much as ～	～ほども (～と同じぐらい多い)	607
349	as low as ～	わずか～ (～と同じほど低い)	542
350	as high as ～	～もの (～と同じほど高い)	491
351	as long as ～	～もの間，～である限り	393
352	as little as ～	わずか～ (～と同じぐらい少ない)	360
353	as effective as ～	～と同じぐらい効果的な	325
354	as many as ～	～もの (～と同じぐらい多い)	291
355	as few as ～	わずか～ (～と同じぐらい少ない)	250
356	as efficiently as ～	～と同じぐらい効果的に	212

346 Frequent injuries caused by falls are seen in sedative abusers **as well as** alcoholics.

- □ **frequent** [3,251]：形 頻繁な
- □ **injury** [15,721]：名 傷害，損傷（類 damage [15,696]，lesion [18,776]）
- □ **caused by ～** [8,510]：～によって引き起こされる
- □ **fall** [2,820]：名 転倒，低下 / 動 低下する
- □ **see** [3,158]：動 見る
- □ **sedative** [162]：形 鎮静の / 名 鎮静剤
- □ **abuser** [110]：名 濫用者
- □ **as well as ～** [26,236]：～と同様に
- □ **alcoholics** [112]：名 アルコール依存患者

訳 転倒によって引き起こされる頻繁な傷害が，アルコール依存患者と同様に鎮静剤の濫用者において見られる

347 This inhibition was dose-dependent, and was observed **as early as** 3 hours after stimulation.

- □ inhibition [37,129]：名 抑制
- □ dose-dependent [4,207]：形 用量依存的な
- □ observe [51,935]：動 観察する
- □ as early as ～ [842]：早くも～
- □ after stimulation [528]：刺激の後

訳 この抑制は用量依存的で，そして刺激の後早くも3時間で観察された

348 The spleens of these mice are **as much as** 30% larger than those of wild-type littermates.

- □ spleen [4,135]：名 脾臓
- □ as much as ～ [607]：～ほども
- □ wild-type [39,564]：野生型の
- □ littermate [1,364]：名 同腹仔

訳 これらのマウスの脾臓は，野生型の同腹仔のそれらより30%ほども大きい

349 The maximum response was observed at a glucose concentration **as low as** 11 mM.

- □ maximum [5,710]：名 最大
- □ response [104,667]：名 応答，反応
- □ observe [51,935]：動 観察する
- □ glucose [16,584]：名 グルコース
- □ concentration [39,910]：名 濃度，集中
- □ as low as ～ [542]：わずか～

訳 最大応答は，わずか11 mMのグルコース濃度で観察された

350 Resistance to this antibiotic in Streptococcus pyogenes was found to be **as high as** 48% in specific populations in the United States.

- □ resistance to ～ [5,635]：～に対する抵抗性
- □ antibiotic [2,279]：名 抗生剤 / 形 抗菌の
- □ Streptococcus pyogenes [201]：化膿性連鎖球菌
- □ found to ～ [15,859]：～することが見つけられる
- □ as high as ～ [491]：～もの
- □ specific [85,296]：形 特異的な
- □ population [28,214]：名 集団，人口

訳 化膿性連鎖球菌におけるこの抗生物質に対する抵抗性は，合衆国における特異的な集団において48％もの高さであることが見つけられた

351 These patients were followed up longitudinally during and after chemotherapy for **as long as** 5 years.

- □ follow up [634]：動 経過観察する
- □ longitudinally [213]：副 長期的に
- □ chemotherapy [5,426]：名 化学療法
- □ as long as ～ [393]：～もの間，～である限り

訳 これらの患者は，化学療法の間および後5年もの間長期的に経過観察された

352 This new method has made it possible to obtain a stack of axial images and reproduce them in **as little as** 15 seconds.

- □ method [32,975]：名 方法
- □ make it possible to ～ [443]：～することを可能にする
- □ obtain [21,689]：動 得る
- □ a stack of ～ [13]：大量の～
- □ axial [1,856]：形 軸方向の
- □ image [20,702]：名 画像 / 動 画像化する

□ **reproduce** [1,486]：動 再現する，生殖する

□ as little as ～ [360]：わずか～

訳 この新しい方法は，大量の軸方向の画像を得て，そしてそれらをわずか15秒で再現することを可能にした

353 These data suggest that androgen treatment may be **as effective as** estrogen replacement in reversing the depression at hippocampal synapses.

□ **suggest that ～** [96,112]：～ということを示唆する

□ **androgen** [5,710]：名 アンドロゲン

□ **treatment** [60,138]：名 治療，処理

□ as effective as ～ [325]：～と同じぐらい効果的な

□ **estrogen replacement** [175]：エストロゲン補充

□ **reverse** [14,176]：動 逆転させる

□ **depression** [5,911]：名 抑圧，うつ，機能低下

□ **hippocampal** [5,410]：形 海馬の

□ **synapse** [6,527]：名 シナプス

訳 これらのデータは，アンドロゲン治療が海馬シナプスにおける抑圧を逆転させる際にエストロゲン補充と同じくらい効果的であるかもしれないということを示唆する

354 **As many as** one in 20 people in the United States have some form of autoimmune disease.

□ as many as ～ [291]：～もの

□ **people** [2,121]：名 人々

□ **form** [9,556]：名 型 / 動 形成する

□ **autoimmune disease** [1,624]：自己免疫疾患

訳 合衆国では20人に1人もの人がなんらかの型の自己免疫疾患を持っている

355 Many of these peptides are very small, comprising **as few as** six residues.

- □ **peptide** [45,549]：名 ペプチド
- □ **small** [27,357]：形 小さい
- □ **comprise** [5,847]：動 構成する
- □ **as few as ～** [250]：わずか～
- □ **residue** [45,837]：名 残基

訳 これらのペプチドの多くは非常に小さく，わずか６残基で構成されている

356 The virus replicated **as efficiently as** wild-type virus in these cells.

- □ **virus** [47,464]：名 ウイルス
- □ **replicate** [3,787]：動 複製する，増殖する（類 proliferate [3,515]，grow [10,566]）
- □ **as efficiently as ～** [212]：～と同じぐらい効果的に
- □ **wild-type** [39,564]：野生型の

訳 そのウイルスは，これらの細胞の中で野生型のウイルスと同じぐらい効果的に複製した

D. as ＋前置詞

文例No			用例数
357	as to ～	～に関して	1,862
358	as for ～	～に関して	844
359	as with ～	～の場合と同じように	892
360	as of ～	～の時点で	305

357 The question arises **as to** whether these physiologically indistinguishable cells have any special functional correlates.

- □ **arise** [6,458]：動 生じる，起こる
- □ **as to ～** [1,862]：～に関して
- □ **physiologically** [1,444]：副 生理学的に
- □ **indistinguishable** [1,683]：形 区別できない
- □ **special** [1,078]：形 特別な
- □ **functional** [34,575]：形 機能的な
- □ **correlate** [21,378]：名 関連現象 / 動 相関させる

訳 これらの生理学的に区別できない細胞が特別な機能的な関連現象をもつかどうかに関して疑問が生じる

358 **As for** gastrointestinal cancer, chemotherapy and radiotherapy have limited efficacy unless the tumor is localized.

- □ **as for ～** [844]：～に関して
- □ **gastrointestinal** [2,243]：形 胃腸の
- □ **chemotherapy** [5,426]：名 化学療法
- □ **radiotherapy** [1,059]：名 放射線療法
- □ **limited** [10,640]：形 限られた
- □ **efficacy** [8,112]：名 有効性
- □ **unless** [816]：～でない限り
- □ **tumor** [61,098]：名 腫瘍

☐ localized [11,504]：局在性の

訳 胃腸癌に関して，腫瘍が局在性でない限り化学療法および放射線療法は限られた有効性しか持たない

359 As with all complex traits, multiple sclerosis is caused by an interplay between unidentified environmental factors and susceptibility genes.

☐ as with ～ [892]：～の場合と同じように
☐ complex [80,256]：形 複雑な / 名 複合体 / 動 複合体を形成する
☐ trait [3,921]：名 形質
☐ multiple sclerosis [1,412]：多発性硬化症
☐ caused by ～ [8,510]：～によって引き起こされる
☐ interplay [787]：名 相互作用
☐ unidentified [1,245]：形 未同定の，未知の（類 unknown [11,802]）
☐ environmental factor [629]：環境因子
☐ susceptibility [7,728]：名 感受性

訳 すべての複雑な形質の場合と同じように，多発性硬化症は未同定の環境因子と感受性遺伝子の間の相互作用によって引き起こされる

360 As of this writing, the novel genome browser application has been set up on the human, mouse, and rat genomes.

☐ as of ～ [305]：～の時点で
☐ writing [123]：名 執筆，記述
☐ novel [30,313]：形 新規の，新しい
☐ genome [22,062]：名 ゲノム
☐ browser [172]：名 ブラウザ
☐ application [10,335]：名 アプリケーション，適用
☐ set up [162]：動 セットアップする

訳 この執筆の時点で，その新規のゲノムブラウザアプリケーションはヒト，マウスおよびラットのゲノムに関してセットアップされている

E. as で始まる節

例文No			用例数
361, 362	as ～	～なので／～であるとき／～につれて／～のように	267,565

361 **As** thallium is excreted in the urine, thallium determinations can be made on 24-hour specimens.

- □ **as** [267,565]：接 ～なので（類 because [28,947], since [11,101]），～であるとき
- □ **thallium** [275]：名 タリウム
- □ **excrete** [290]：動 排泄する，排出する
- □ **urine** [2,266]：名 尿
- □ **determination** [4,378]：名 定量，決定
- □ **specimen** [5,996]：名 検体

訳 タリウムは尿に排泄されるので，タリウム定量は24時間検体に対してなされうる

362 Particularly difficult times can arise **as** physicians begin dealing with patients who are becoming old, frail, or cognitively impaired.

- □ **particularly** [7,227]：副 特に（類 especially [4,892], notably [1,715], in particular [3,662]）
- □ **difficult** [3,414]：形 困難な，難しい（類 hard [530]）
- □ **as** [267,565]：接 ～であるとき（類 when [59,392]），～なので
- □ **arise** [6,458]：動 起こる，生じる
- □ **physician** [4,789]：名 内科医，医師
- □ **begin** [3,922]：動 始める，始まる
- □ **deal with ～** [338]：～を扱う
- □ **become** [12,204]：動 ～になる
- □ **old** [5,600]：形 年老いた（類 elderly [2,100]），古い
- □ **frail** [37]：形 虚弱な（類 weak [3,617]）

□ **cognitively** [119]：副 認知的に，認識的に

□ **impair** [10,580]：動 障害する，損なう

訳 内科医が，年老い，虚弱になり，あるいは認知障害された患者を扱い始めるとき，特に困難な時期が起こりうる

Ⅶ. 熟語表現

熟語とは，いくつかの単語を組み合わせることによって本来の意味とは少し異なる内容を表すもので，英語を使う上で非常に重要な表現法である．ここでは論文でよく用いられる熟語／熟語的表現について取り上げる．

熟語表現の文例

文例No			用例数
363	in the absence of ～	～なしで	12,670
364	after adjusting for ～	～に対して調整した後	550
211	after adjustment for ～	～について調整した後	1,359
365	in good agreement with ～	～とよく一致している	306
235	in association with ～	～と関連して	599
366	in an attempt to ～	～しようとして	759
367	on the basis of ～	～に基づいて	4,874
368	at the beginning of ～	～の初めに	228
369	in the case of ～	～の場合に	1,411
370	in combination with ～	～と組合わせて	2,878
－	when combined with ～	～と組合わせると	404
371	in common with ～	～と共通して	219
372	in complex with ～	～と複合した	885
373	at a concentration of ～	～の濃度で	208
－	in concert with ～	～と協調して	282
374	under conditions of ～	～の条件下で	767
375	under the same conditions	同じ条件下で	290
376	in conjunction with ～	～と組合わせて	2,007
377	in contact with ～	～と接触して	322

文例No			用例数
378	in the context of ～	～との関連で	2,645
379	within the context of ～	～という脈略のなかで	273
380	under the control of ～	～の制御下で	1,284
–	after controlling for ～	～に対して調節した後で	155
381	during the course of ～	～の経過の間に	608
–	in the course of ～	～の経過において	260
382	over the course of ～	～の経過の間に	348
383	at a dose of ～	～の用量で	321
384	in an effort to ～	～しようとして	890
385	at the end of ～	～の終わりに	1,189
386	to this end	この目的のために	388
387	lines of evidence	一連の証拠	694
388	for example	たとえば	2,554
389	in excess of ～	～を超えて	280
390	in the face of ～	～にもかかわらず，～に直面して	321
391	in favor of ～	～を支持する	348
392	in the form of ～	～の形で	785
393	give rise to ～	～を生じる	2,504
394	for instance	たとえば	223
395	to our knowledge	我々の知る限りでは	857
248	little is known about ～	～についてはほとんど知られていない	3,195
396	in light of ～	～に照らして	602
397	shed light on ～	～の解明に役立つ，～を明らかにする	481
398	in a manner similar to ～	～に類似した様式で	391
399	by means of ～	～を用いて	1,569
400	on the order of ～	～のオーダーで	267
194	in order to ～	～するために	4,148

文例No			用例数
401	at least in part	少なくとも一部は	1,713
402	in place of ～	～の代わりに	529
403	in the presence of ～	～の存在下で	15,708
404	during the process of ～	～の過程の間に	184
405	in proportion to ～	～に比例して	198
406	in close proximity to ～	～のすぐ近くに	412
407	for the purpose of ～	～の目的で	201
105	raise the possibility that ～	～という可能性を示唆する	298
408	raise the question of ～	～の疑問を提起する	239
409	at a rate of ～	～の割合で，～の速度で	284
410	with regard to ～	～に関して	1,048
189	with respect to ～	～に関して	2,069
265	in response to ～	～に応答して	8,464
－	at risk for ～	～の危険がある	629
399	at high risk for ～	～のリスクの高い	367
411	in support of ～	～を支持して	715
－	take advantage of ～	～を利用する	88
－	take into account ～	～を考慮に入れる	68
412	take place	起こる	1,287
253	in terms of ～	～に関して	3,377
413	for the first time	はじめて	2,700
－	at the same time	同時に	341
414	in view of ～	～を考慮して	389
－	by virtue of ～	～のおかげで	429
415	by use of ～	～の使用によって	1,116

363 This mechanism of DNA binding may play an important role in maintaining stable negative regulation of this particular gene expression **in the absence of** extracellular stimulation.

- □ DNA binding [11,354]：DNA 結合
- □ play an important role in ～ [6,552]：～の際に重要な役割を果たす
- □ maintain [13,333]：動 維持する
- □ stable [13,642]：形 安定な
- □ negative [25,102]：形 負の
- □ regulation [35,679]：名 調節，制御（類 modulation [5,966]）
- □ particular [7,937]：形 特定の
- □ gene expression [21,171]：遺伝子発現
- □ in the absence of ～ [12,670]：～なしで（反 in the presence of [15,708]）
- □ extracellular [16,514]：形 細胞外の（反 intracellular [20,566]）
- □ stimulation [21,013]：名 刺激

訳 このDNA結合の機構は，細胞外刺激なしでこの特定の遺伝子発現の安定した負の制御を維持する際に重要な役割を果たすかもしれない

364 **After adjusting for** body mass index, age, and sex, there was no significant difference in the serum cholesterol concentrations.

- □ after adjusting for ～ [550]：～に対して調整した後
- □ body mass index [1,523]：ボディマス指数，肥満度指数
- □ significant [43,571]：形 有意な，著しい，重要な
- □ difference in ～ [16,401]：～の違い
- □ serum cholesterol [282]：血清コレステロール
- □ concentration [39,910]：名 濃度，集中

訳 ボディマス指数，年齢，性別に対して調整した後，血清コレステロール濃度の有意な違いはなかった

365 Our results obtained with the molecular dynamics simulation technique are **in good agreement with** the data available from experiments.

- □ **obtain** [21,689]：動 得る
- □ **dynamics** [7,155]：名 動力学
  molecular dynamics [1,372]：分子動力学
- □ **simulation** [5,095]：名 シミュレーション
- □ **technique** [14,941]：名 技術，技法，テクニック
- □ **in good agreement with ～** [306]：～とよく一致している
- □ **available** [10,780]：形 入手できる，利用できる
- □ **experiment** [23,342]：名 実験

訳 分子動力学シミュレーション技術によって得られた我々の結果は，実験から入手できる結果とよく一致している

366 **In an attempt to** circumvent this adverse effect, we evaluated the efficacy of aerosol and oral vaccinations.

- □ **in an attempt to ～** [759]：～しようとして
- □ **circumvent** [566]：動 回避する（類 avoid [2,051]）
- □ **adverse effect** [1,014]：有害作用，副作用
- □ **evaluate** [19,786]：動 評価する
- □ **efficacy** [8,112]：名 有効性
- □ **aerosol** [728]：名 エアロゾル
- □ **oral** [6,343]：形 経口の，口腔の
- □ **vaccination** [2,988]：名 ワクチン接種

訳 この有害作用を回避しようとして，我々はエアロゾルワクチン接種および経口ワクチン接種の有効性を評価した

367 **On the basis of** these findings, we propose that tyrosine phosphorylation is not involved in the signal transduction pathway leading to cell growth suppression.

- □ **on the basis of ～** [4,874]：～に基づいて

- □ finding [35,256]：名 知見
- □ we propose that ～ [6,260]：我々は～ということを提唱する
- □ tyrosine [19,801]：名 チロシン
- □ phosphorylation [34,581]：名 リン酸化
- □ involved in ～ [25,574]：～に関与する
- □ signal transduction pathway [2,354]：シグナル伝達経路
- □ lead to ～ [31,651]：～につながる
- □ cell growth [4,368]：細胞増殖
- □ suppression [7,483]：名 抑制

訳 これらの知見に基づいて，我々はチロシンリン酸化が細胞増殖抑制につながるシグナル伝達系路に関与しないということを提唱する

VII 熟語表現

368 Blood pressure and peripheral vascular resistance fell **at the beginning of** pregnancy.

- □ blood pressure [3,723]：血圧
- □ peripheral vascular resistance [23]：末梢血管抵抗
- □ fall [2,820]：動（過 fell）低下する（類 decrease [51,139], decline [6,411]）/ 名 転倒，低下
- □ at the beginning of ～ [228]：～の初めに
- □ pregnancy [3,407]：名 妊娠

訳 妊娠の初めに，血圧および末梢血管抵抗が低下した

369 **In the case of** rare hereditary disease, a careful family history is indispensable in assessing and understanding the disease.

- □ in the case of ～ [1,411]：～の場合に
- □ rare [3,385]：形 稀な
- □ hereditary [1,428]：形 遺伝性の
- □ disease [71,437]：名 疾患
- □ careful [568]：形 念入りな，慎重な，注意深い
- □ family history [867]：家族歴
- □ indispensable [377]：形 不可欠な

□ **assess** [20,743]：動 評価する

□ **understand** [12,582]：動 理解する

訳 稀な遺伝性疾患の場合に，念入りな家族歴が疾患を評価して理解する際に不可欠である

370

Congenital forms of ACTH deficiency usually occur **in combination with** the loss of other pituitary hormones.

□ **congenital** [1,980]：形 先天性の

□ **form** [9,556]：名 型 / 動 形成する

□ **ACTH** [493]：名 副腎皮質刺激ホルモン

□ **deficiency** [6,979]：名 欠損，欠損症

□ **usually** [3,166]：副 普通，通常

□ **occur** [42,905]：動 起こる，生じる

□ **in combination with ～** [2,878]：～と組合わせて

□ **loss** [28,751]：名 喪失，減少

□ **pituitary** [1,963]：形 下垂体の

□ **hormone** [10,499]：名 ホルモン

訳 先天性の型の副腎皮質刺激ホルモン欠損症は，通常，他の下垂体ホルモンの喪失と組合わせて起こる

371

The estrogen receptor, **in common with** other nuclear hormone receptors, contains two transcription activation functions.

□ **estrogen receptor** [1,835]：エストロゲン受容体

□ **in common with ～** [219]：～共通して

□ **nuclear hormone receptor** [369]：核内ホルモン受容体

□ **contain** [71,403]：動 含む

□ **transcription** [51,307]：名 転写

□ **activation** [82,818]：名 活性化

□ **function** [92,343]：名 機能 / 動 機能する

訳 エストロゲン受容体は，他の核内ホルモン受容体と共通して，２つの転写活性化機能を含む

372 We analyzed the crystal structures of other bone morphogenetic proteins **in complex with** their receptors.

- □ **analyze** [18,249]：動 分析する，解析する
- □ **crystal structure** [6,231]：結晶構造
- □ **bone morphogenetic protein** [663]：骨形成タンパク質
- □ **in complex with ～** [885]：～と複合した
- □ **receptor** [125,789]：名 受容体

訳 我々は，それらの受容体と複合した他の骨形成タンパク質の結晶構造を分析した

373 Without exception, these compounds were all inactive **at a concentration of** 10 μM.

- □ **exception** [1,696]：名 例外
- □ **compound** [16,256]：名 化合物
- □ **inactive** [5,423]：形 不活性の
- □ **at a concentration of ～** [208]：～の濃度で

訳 例外なしに，これらの化合物は10μMの濃度ですべて不活性であった

374 Further studies are needed to assess the utility of highly selective inhibitors of glycogen synthase kinase-3 for the modification of insulin action **under conditions of** insulin resistance.

- □ **further** [29,238]：形 さらに進んだ / 副 さらに
- □ **need** [12,039]：動 必要とする / 名 必要性
- □ **assess** [20,743]：動 評価する
- □ **utility** [3,086]：名 有用性（類 usefulness [703]）
- □ **selective** [15,241]：形 選択的な
- □ **inhibitor** [43,673]：名 阻害剤
- □ **glycogen synthase kinase** [454]：グリコーゲン合成酵素キナーゼ
- □ **modification** [9,701]：名 修飾
- □ **insulin** [20,353]：名 インスリン

☐ **action** [17,526]：名 作用

☐ **under conditions of ～** [767]：～の条件下で

☐ **resistance** [20,333]：名 抵抗性

訳 さらなる研究が，インスリン抵抗性の条件下でのインスリン作用の修飾に対するグリコーゲン合成酵素キナーゼ3の高度に選択的な阻害剤の有用性を評価するために必要とされる

375 **Under the same conditions**, cooperativity was much greater for the pancreatic form of the enzyme than for the liver form.

☐ **under the same conditions** [290]：同じ条件下で

☐ **cooperativity** [1,354]：名 協同性

☐ **much greater** [485]：ずっと大きい

☐ **pancreatic** [5,040]：形 膵臓の

☐ **form** [9,556]：名 型 / 動 形成する

☐ **enzyme** [49,164]：名 酵素

訳 同じ条件下で，その酵素の膵臓型に対する協同性は肝臓型に対してよりもずっと大きかった

376 Iritis is known to occur frequently **in conjunction with** intestinal inflammation.

☐ **iritis** [10]：名 虹彩炎

☐ **known to ～** [8,375]：～すると知られている

☐ **occur** [42,905]：動 起こる，生じる

☐ **frequently** [5,561]：副 しばしば

☐ **in conjunction with ～** [2,007]：～と組合わせて

☐ **intestinal** [6,063]：形 小腸の

☐ **inflammation** [9,268]：名 炎症

訳 虹彩炎は，しばしば小腸炎と組合わせて起こることが知られている

377 The mitral and aortic valves are normally **in contact with** each other.

- □ mitral [1,447]：形 僧帽弁の
- □ aortic [3,978]：形 大動脈の
- □ valve [2,269]：名 弁
- □ normally [6,493]：副 通常は
- □ in contact with ～ [322]：～と接触して
- □ each other [2,962]：お互いに

訳 僧帽弁と大動脈弁は，通常はお互いに接触している

378 We discuss our findings **in the context of** mechanisms that may underlie tumorigenesis.

- □ discuss [9,273]：動 議論する
- □ finding [35,256]：名 知見
- □ in the context of ～ [2,645]：～との関連で
- □ underlie [10,696]：動 根底にある
- □ tumorigenesis [2,442]：名 腫瘍形成，腫瘍発生

訳 我々は，腫瘍形成の根底にあるかもしれない機構との関連で我々の知見を議論する

379 We have established physiological roles and sites of action of the different topoisomerases **within the context of** the bacterial cell cycle.

- □ establish [19,016]：動 確立する
- □ physiological [9,754]：形 生理的な，生理学的な
- □ role [87,150]：名 役割
- □ site [99,321]：名 部位
- □ action [17,526]：名 作用
- □ different [48,428]：形 異なる
- □ topoisomerase [2,462]：名 トポイソメラーゼ
- □ within the context of ～ [273]：～という脈絡のなかで

☐ **bacterial** [13,176]：形 細菌の

☐ **cell cycle** [13,210]：細胞周期

訳 我々は，細菌の細胞周期という脈絡のなかで異なるトポイソメラーゼの生理的な役割と作用の部位を確立した

380 Each of these transgenes is **under the control of** individual poxvirus promoters.

☐ **transgene** [5,659]：名 導入遺伝子

☐ **under the control of ～** [1,284]：～の制御下で

☐ **individual** [24,178]：形 個々の / 名 個人

☐ **poxvirus** [252]：名 ポックスウイルス

☐ **promoter** [40,801]：名 プロモーター

訳 これらの導入遺伝子のおのおのは，個々のポックスウイルスプロモーターの制御下にある

381 Response to this drug varies considerably among patients **during the course of** therapy.

☐ **response to ～** [36,619]：～の応答

☐ **vary** [9,676]：動 変動する

☐ **considerably** [1,606]：副 かなり（類 substantially [4,692]，quite [1,408]）

☐ **during the course of ～** [608]：～の過程の間に

☐ **therapy** [28,037]：名 治療，療法

訳 この薬剤への応答は，治療の過程の間に患者の間でかなり変動する

382 Here, we present evidence that this family of transposable elements has been significantly amplified **over the course of** evolution.

☐ **present evidence** [1,433]：証拠を提示する

☐ **transposable** [486]：形 転位性の

transposable element [460]：転位因子

☐ **significantly** [48,939]：副 顕著に，有意に

□ amplify [3,855]：動 増幅する

□ over the course of ～ [348]：～の過程の間に

□ evolution [7,907]：名 進化

訳 ここに我々は，このファミリーの転位因子は進化の過程の間に顕著に増幅されてきたという証拠を示す

383 This anticancer chemotherapy drug was administered **at a dose of** 1 mg/kg per day for 3 days.

□ anticancer [1,282]：形 抗癌性の

□ chemotherapy [5,426]：名 化学療法

□ administer [6,478]：動 投与する，投薬する

□ at a dose of ～ [321]：～の用量で

□～ per day [1,100]：1日につき～

訳 この抗癌化学療法剤が，1日あたり1 mg/kgの用量で3日間投与された

384 **In an effort to** characterize mechanisms common to these processes, we describe the earliest stages of offspring formation in this prokaryote.

□ in an effort to ～ [890]：～しようとして

□ characterize [27,658]：動 特徴づける

□ common to ～ [1,006]：～に共通する

□ process [50,139]：名 過程 / 動 処理する

□ describe [25,975]：動 述べる

□ earliest [1,331]：形 最も早い

□ stage [19,952]：名 期，ステージ / 動 段階に分ける

□ offspring [1,700]：名 子孫

□ formation [41,324]：名 形成

□ prokaryote [677]：名 原核生物

訳 これらの過程に共通する機構を特徴づけしようとして，我々はこの原核生物における子孫形成の最も早いステージについて述べる

385 **At the end of** the experiment, the dystrophic rats were perfused and processed for histologic assessment of photoreceptor survival.

- □ at the end of ～ [1,189]：～の終わりに
- □ experiment [23,342]：名 実験
- □ dystrophic [310]：形 ジストロフィーの
- □ perfuse [1,395]：動 灌流させる
- □ process [50,139]：動 処理する / 名 過程
- □ histologic [2,062]：形 組織の
- □ assessment [5,256]：名 評価
- □ photoreceptor [3,736]：名 光受容体
- □ survival [28,139]：名 生存，生存率

訳 実験の終わりに，ジストロフィーのラットは光受容体生存の組織評価のために灌流されそして処理された

386 **To this end**, we have developed a novel method for the differential enhancement of probe sequence concentration by subtractive hybridization.

- □ to this end [388]：この目的のために
- □ develop [38,325]：動 開発する，発症する，発達する
- □ novel [30,313]：形 新規の，新しい
- □ method for ～ [4,979]：～のための方法
- □ differential [7,310]：形 差動的な
- □ enhancement [5,207]：名 増強
- □ probe [12,160]：名 プローブ / 動 探索する
- □ sequence [91,510]：名 配列 / 動 配列決定する
- □ concentration [39,910]：名 濃度，集中
- □ subtractive [267]：形 減算の，サブトラクティブな
- □ hybridization [7,179]：名 ハイブリダイゼーション
- □ subtractive hybridization [176]：サブトラクティブハイブリダイゼーション（組織特異的に発現する遺伝子を同定する方法）

訳 この目的のために，我々はサブトラクティブハイブリダイゼーションによるプローブ配列濃度の異なる増強のための新規の方法を開発した

387 These data contribute to several **lines of evidence** indicating that genetically distinct subtypes of motor neurons are specified during development.

- □ contribute to ～ [19,994]：～に寄与する
- □ **lines of evidence** [694]：一連の証拠
- □ indicate that ～ [58,147]：～ということを示す
- □ genetically [4,530]：副 遺伝的に
- □ distinct [22,633]：形 別個の，明らかに異なる
- □ subtype [5,629]：名 サブタイプ
- □ motor [10,040]：形 運動の / 名 モーター
- □ neuron [39,544]：名 ニューロン
- □ specify [2,532]：動 特定化する
- □ development [48,509]：名 発生，発症，開発

訳 これらのデータは，遺伝的に別個のサブタイプの運動神経が発生の間に特定化されるということを示すいくつかの一連の証拠に寄与する

388 **For example**, two recently published reviews investigated the evidence on racial/ethnic differences in medical care.

- □ **for example** [2,554]：たとえば
- □ recently [15,473]：副 最近（類 previously [34,828]）
- □ publish [3,595]：動 出版する，発表する
- □ review [12,123]：名 総説，概説 / 動 概説する
- □ investigate [30,619]：動 精査する
- □ evidence [35,499]：名 証拠 / 動 立証する
- □ racial [546]：形 人種の
- □ ethnic [973]：形 民族の
- □ difference in ～ [16,401]：～の違い
- □ medical [6,547]：形 医学の
- □ medical care [406]：医療

VII 熟語表現

訳 たとえば，2つの最近出版された総説は医療における人種／民族の違いに関する証拠を精査した

389 The estimated annual hospital cost is **in excess of** $7 billion.

- □ estimate [13,499]：動 見積もる
- □ annual [1,178]：形 年間の，年一回の
- □ hospital [7,176]：名 病院
- □ cost [7,217]：名 コスト，費用
- □ in excess of ～ [280]：～を超えて
- □ billion [446]：名 十億

訳 予想される年間の病院のコストは，70億ドルを超える

390 This may help to preserve the function of the retinal pigment epithelium **in the face of** aging and disease.

- □ help to ～ [1,718]：～するのに役立つ
- □ preserve [2,565]：動 保存する
- □ function [92,343]：名 機能 / 動 機能する
- □ retinal pigment epithelium [266]：網膜色素上皮
- □ in the face of ～ [321]：～にもかかわらず，～に直面して
- □ aging [2,738]：名 加齢，老化
- □ disease [71,437]：名 疾患

訳 これは，加齢と疾患にもかかわらず網膜色素上皮の機能を保存するのに役立つかもしれない

391 These findings constitute strong evidence **in favor of** the beneficial effects of glucosamine.

- □ finding [35,256]：名 知見
- □ constitute [3,343]：動 構成する
- □ evidence [35,499]：名 証拠 / 動 立証する
- □ in favor of ～ [348]：～を支持する

- □ **beneficial** [2,305]：形 有益な
- □ **effect** [106,593]：名 効果，影響
- □ **glucosamine** [437]：名 グルコサミン

訳 これらの知見は，グルコサミンの有益な効果を支持する強力な証拠を構成する

392 Ischemic injury to the kidney is characterized in part by tubular cell death **in the form of** necrosis or apoptosis.

- □ **ischemic injury** [349]：虚血傷害
- □ **kidney** [10,654]：名 腎臓
- □ **characterize** [27,658]：動 特徴づける
- □ **in part** [6,718]：一部は，部分的に
- □ **tubular cell** [179]：尿細管細胞
- □ **cell death** [9,708]：細胞死
- □ **in the form of ～** [785]：～の形で
- □ **necrosis** [7,242]：名 ネクローシス，壊死
- □ **apoptosis** [31,507]：名 アポトーシス

訳 腎臓に対する虚血傷害は，ネクローシスあるいはアポトーシスの形で尿細管細胞死によって一部は特徴づけられる

393 The tetraploid cells **gave rise to** malignant mammary epithelial cancers when transplanted subcutaneously into nude mice.

- □ **tetraploid** [235]：名 四倍体
- □ **give rise to ～** [2,504]：～を生じる
- □ **malignant** [4,511]：形 悪性の
- □ **mammary** [4,405]：形 乳腺の，乳房の
- □ **epithelial** [16,204]：形 上皮の
- □ **transplant** [10,234]：動 移植する / 名 移植片
- □ **subcutaneously** [572]：副 皮下に
- □ **nude mouse** [1,322]：ヌードマウス

訳 四倍体細胞は，ヌードマウスに皮下に移植されたとき，悪性の乳腺上皮癌を生じた

394 **For instance**, in some forms of leukemia, younger patients respond to therapy better than older patients.

- □ for instance ［223］：たとえば
- □ form ［9,556］：名 型 / 動 形成する
- □ leukemia ［7,767］：名 白血病
- □ younger ［2,136］：形 より若い
- □ respond to ～ ［5,457］：～に反応する，～に応答する
- □ therapy ［28,037］：名 治療，療法
- □ better than ～ ［1,089］：～よりよい
- □ older ［4,766］：より高齢の

訳 たとえば，いくつかの型の白血病において，より若い患者は高齢の患者より治療によく反応する

395 **To our knowledge**, these results provide the first evidence that haplotypes in the growth hormone secretagogue receptor gene are involved in the pathogenesis of human obesity.

- □ to our knowledge ［857］：我々の知る限りでは
- □ provide ［56,723］：動 提供する，与える
- □ evidence that ～ ［9,965］：～という証拠
- □ haplotype ［3,386］：名 ハプロタイプ
- □ growth hormone secretagogue ［25］：ホルモン分泌促進因子
- □ receptor ［125,789］：名 受容体
- □ involved in ～ ［25,574］：～に関与する
- □ pathogenesis ［8,673］：名 病因
- □ obesity ［3,240］：名 肥満

訳 我々の知る限りでは，これらの結果は成長ホルモン分泌促進因子受容体遺伝子のハプロタイプはヒトの肥満の病因に関与するという最初の証拠を提供する

396 **In light of** recent studies, it is now clear that mutations outside the kinase domain can lead to enhanced autophosphorylation of the kinase.

- ☐ **in light of ～** [602]：～に照らして
- ☐ recent [18,452]：形 最近の
- ☐ clear [5,786]：形 明らかな（類 apparent [8,392], evident [2,899], obvious [1,088], distinct [22,633]）/ 動 除去する
- ☐ mutation [70,231]：名 変異
- ☐ kinase [68,918]：名 キナーゼ，リン酸化酵素
- ☐ domain [87,363]：名 ドメイン
- ☐ lead to ～ [31,651]：～につながる
- ☐ enhanced [20,161]：増強した
- ☐ autophosphorylation [1,603]：名 自己リン酸化

訳 最近の研究に照らして，キナーゼドメインの外側の変異はそのキナーゼの増強した自己リン酸化につながりうるということが今明らかである

397 These findings **shed light on** the molecular function and evolution of clock genes in vertebrates.

- ☐ finding [35,256]：名 知見
- ☐ **shed light on ～** [481]：～の解明に役立つ，～を明らかにする
- ☐ molecular function [132]：分子機能
- ☐ evolution [7,907]：名 進化
- ☐ clock [1,911]：名 時計
- ☐ vertebrate [6,044]：名 脊椎動物

訳 これらの知見は，脊椎動物における時計遺伝子の分子機能と進化の解明に役立つ

398 Physicians should think of anemia **in a manner similar to** the way in which they regard symptoms like diarrhea as an indicator or a manifestation of an underlying disease.

- ☐ physician [4,789]：名 内科医，医師

- □ **think of ～** [107]：～を考える
- □ **in a manner similar to ～** [391]：～に類似した様式で
- □ **way** [5,601]：名 方法
- □ **regard** [3,713]：動 みなす，考える
- □ **symptom** [9,302]：名 症状
- □ **diarrhea** [1,375]：名 下痢
- □ **indicator** [2,268]：名 指標
- □ **manifestation** [1,760]：名 徴候，症状
- □ **underlying** [7,940]：根底にある
- □ **disease** [71,437]：名 疾患

訳 内科医は，彼らが下痢のような症状を根底にある疾患の指標あるいは徴候とみなす方法に似た様式で貧血を考えるべきである

399 **By means of** a questionnaire, we identified women at high risk for chlamydial infection.

- □ **by means of ～** [1,569]：～を用いて
- □ **questionnaire** [2,089]：名 アンケート
- □ **identify** [73,456]：動 同定する
- □ **at high risk for ～** [369]：～のリスクの高い
- □ **chlamydial** [690]：形 クラミジアの
- □ **infection** [44,601]：名 感染

訳 アンケートを用いて，我々はクラミジア感染症のリスクの高い女性を同定した

400 The influenza vaccine was successful in protecting against pneumonia and death **on the order of** 60 to 80%.

- □ **influenza** [2,999]：名 インフルエンザ
- □ **vaccine** [10,085]：名 ワクチン
- □ **successful** [3,728]：形 成功した
- □ **protect against ～** [1,800]：～から保護する
- □ **pneumonia** [1,737]：名 肺炎
- □ **death** [26,658]：名 死

☐ on the order of ～ [267]：～のオーダーで

訳 そのインフルエンザワクチンは，60～80％のオーダーで肺炎および死から保護することに成功した

401 Rates of cervical cancer have declined **at least in part** due to cytologic screening for cervical pathology.

☐ rate [59,576]：名 割合，速度

☐ cervical [2,154]：形 頸部の，子宮頸部の

cervical cancer [428]：子宮頸癌

☐ decline [6,411]：動 低下する（類 decrease [51,139]），名 低下

☐ at least in part [1,713]：少なくとも一部は

☐ due to ～ [18,077]：～のせいで，～のゆえに

☐ cytologic [122]：形 細胞学的な

☐ screening [3,604]：名 スクリーニング

☐ pathology [3,222]：名 病理，病理学，病態

訳 子宮頸部癌の割合は，少なくとも一部は頸部病理の細胞学的なスクリーニングゆえに低下してきた

402 A number of alternative antibiotics can be used **in place of** penicillin.

☐ a number of ～ [7,385]：いくつかの～

☐ alternative [8,904]：形 代替の / 名 代替物

☐ antibiotics [1,694]：名 抗生剤，抗生物質

☐ in place of ～ [529]：～の代わりに

☐ penicillin [653]：ペニシリン

訳 いくつかの代替の抗生剤が，ペニシリンの代わりに使われうる

403 Many clinical decisions require careful dose adjustment of drugs **in the presence of** renal dysfunction.

☐ clinical [31,256]：形 臨床の

☐ decision [2,859]：名 決定

VII 熟語表現

- □ **require** [65,968]：動 必要とする
- □ **careful** [568]：形 慎重な，注意深い，念入りな
- □ **adjustment** [2,964]：名 調整
- □ **in the presence of 〜** [15,708]：〜の存在下で
- □ **renal dysfunction** [229]：腎機能障害

訳 多くの臨床的な決定は，腎機能障害の存在下で薬剤の慎重な用量調整を必要とする

404 These modifications of cellular gene expression probably occur **during the process of** liver cell injury and repair.

- □ **modification** [9,701]：名 修飾
- □ **cellular** [29,968]：形 細胞の，細胞性の
- □ **gene expression** [21,208]：遺伝子発現
- □ **probably** [5,067]：副 おそらく，ありそうに
- □ **occur** [42,905]：動 起こる，生じる
- □ **during the process of 〜** [184]：〜の過程の間に
- □ **injury** [15,721]：名 障害，傷害
- □ **repair** [13,402]：名 修復 / 動 修復する

訳 細胞の遺伝子発現のこれらの修飾は，肝臓細胞障害と修復の過程の間におそらく起こる

405 Increasing evidence demonstrates that risks of lung cancer rise **in proportion to** both duration of smoking and amount smoked per day.

- □ **increasing** [11,608]：形 増大する
- □ **evidence** [35,499]：名 証拠 / 動 立証する
- □ **demonstrate that 〜** [45,052]：〜ということを実証する
- □ **risk** [37,608]：名 リスク，危険
- □ **lung cancer** [2,400]：肺癌
- □ **in proportion to 〜** [198]：〜に比例して
- □ **duration** [7,192]：名 持続時間
- □ **smoking** [3,471]：名 喫煙

☐ amount [11,659]：名 量

☐ smoke [1,024]：動 喫煙する

☐～ per day [1,100]：1日につき～

訳 増大する証拠は，肺癌のリスクが喫煙期間と1日あたりの喫煙量の両方に比例して上がるということを実証する

406 Shortly after the infection, the microorganisms were found in the submucosa **in close proximity to** local immune cells and blood vessels.

☐ shortly [716]：副 直ちに

shortly after ～ [573]：～のあと直ちに

☐ infection [44,601]：名 感染

☐ microorganism [1,151]：名 微生物

☐ submucosa [64]：名 粘膜下層

☐ **in close proximity to ～** [412]：～のすぐ近くに

☐ local [9,593]：形 局所の

☐ immune cell [837]：免疫細胞

☐ blood vessel [837]：血管

訳 感染のあと直ちに，それらの微生物が局所の免疫細胞と血管のすぐ近くの粘膜下層に見つけられた

407 Some victims of accidents or work injuries may exaggerate their pain **for the purpose of** compensation or for psychological reasons.

☐ victim [277]：名 被害者

☐ accident [223]：名 事故，アクシデント

☐ injury [15,721]：名 傷害，損傷

☐ exaggerate [503]：動 誇張する

☐ pain [5,403]：名 痛み

☐ **for the purpose of ～** [201]：～の目的で

☐ compensation [763]：名 補償

☐ psychological [794]：形 心理的な，心理学的な

□ **reason** [2,180]：名 理由 / 動 判断する

訳 事故あるいは仕事上の傷害の被害者は，補償の目的あるいは心理的な理由で彼らの苦痛を誇張するかもしれない

408 Purulent sputum **raises the question of** chronic bronchitis, bronchiectasis, pneumonia, or lung abscess.

□ **purulent** [28]：形 化膿性の

□ **sputum** [498]：名 痰

□ **raise the question of ～** [239]：～の疑問を提起する

□ **chronic** [14,659]：形 慢性の

□ **bronchitis** [163]：名 気管支炎

□ **bronchiectasis** [58]：名 気管支拡張症

□ **pneumonia** [1,737]：名 肺炎

□ **abscess** [498]：名 膿瘍

訳 化膿性の痰は，慢性気管支炎，気管支拡張症，肺炎，肺膿瘍の疑問を提起する

409 The size of the treated population has been expanding **at a rate of** 5% per year.

□ **at a rate of ～** [284]：～の割合で，～の速度で

□ **size** [18,211]：名 大きさ，サイズ

□ **treat** [29,577]：動 治療する，処理する

□ **population** [28,214]：名 集団，人口

□ **expand** [4,692]：動 広がる

□ **per year** [894]：１年につき

訳 治療される集団の大きさは，年５%の割合で広がってきている

410 **With regard to** adult stem cells, their use in research and therapy is not controversial because the production of adult stem cells does not require the destruction of an embryo.

- □ with regard to ～ [1,048]：～に関して
- □ stem cell [7,364]：幹細胞
- □ use in ～ [1,048]：～における使用
- □ research [8,332]：名 研究，調査
- □ therapy [28,037]：名 治療，療法
- □ controversial [1,630]：形 議論の的になる
- □ production [27,611]：名 産生
- □ require [65,968]：動 必要とする
- □ destruction [2,152]：名 破壊
- □ embryo [11,503]：名 胚

訳 成体幹細胞に関して，成体幹細胞の産生は胚の破壊を必要としないので，研究と治療におけるそれらの使用は議論の的にならない

411 **In support of** this hypothesis, we demonstrate that both of these phenotypes are strain dependent.

- □ in support of ～ [715]：～を支持して
- □ hypothesis [15,247]：名 仮説
- □ demonstrate that ～ [45,052]：～ということを実証する
- □ phenotype [23,661]：名 表現型
- □ strain [33,374]：名 菌株，系統
- □ dependent [70,633]：形 依存性の

訳 この仮説を支持して，我々はこれらの表現型の両方が菌株依存性であることを実証する

412 The gene rearrangements **took place** at various stages of the differentiation pathway.

- □ rearrangement [4,914]：名 再編成

Ⅶ 熟語表現

- ☐ take place ［1,287］：起こる
- ☐ various ［14,339］：形 さまざまな
- ☐ stage ［19,952］：名 期，ステージ / 動 段階に分ける
- ☐ differentiation pathway ［125］：分化経路

訳 遺伝子再編成は，分化経路のさまざまなステージで起こった

413 These data show **for the first time** that exercise performed immediately following immobilization induces profound changes in the expression of a number of genes in favor of the restoration of muscle mass.

- ☐ show that ～ ［74,382］：～ということを示す
- ☐ for the first time ［2,700］：はじめて
- ☐ exercise ［5,249］：名 運動／動 運動する
- ☐ perform ［21,488］：動 行う，実行する
- ☐ immediately ［3,602］：副 すぐに
- ☐ follow ［44,620］：動 続く
- ☐ immobilization ［444］：名 固定化，不動化
- ☐ induce ［116,671］：動 誘導する
- ☐ profound ［2,293］：形 著明な，深刻な
- ☐ change in ～ ［30,981］：～の変化
- ☐ expression ［154,475］：名 発現
- ☐ a number of ～ ［7,385］：いくつかの～
- ☐ in favor of ～ ［348］：～を支持する
- ☐ restoration ［1,525］：名 回復，修復
- ☐ muscle mass ［216］：筋量

訳 これらのデータは，固定化の直後に行われる運動は筋量の回復を支持するいくつかの遺伝子の発現の著明な変化を誘導するということをはじめて示す

414 **In view of** their potential for malignancy, it is generally recommended that polyps of any size causing significant symptoms be removed.

- □ in view of ～ [389]：～を考慮して
- □ potential [37,200]：名 可能性，潜在能，形 潜在的な，可能な
- □ malignancy [3,123]：名 悪性度，悪性病変
- □ generally [5,546]：副 一般に
- □ recommend [1,995]：動 勧める
- □ polyp [875]：名 ポリープ
- □ size [18,211]：名 大きさ，サイズ
- □ cause [46,697]：動 引き起こす / 名 原因
- □ significant [43,571]：形 著しい，重要な，有意な
- □ symptom [9,302]：名 症状
- □ remove [5,465]：動 切除する，除去する

訳 それらの悪性度に対する可能性を考慮して，著しい症状を引き起こすどの大きさのポリープも切除されることが一般に勧められる

415 The development of resistance may be delayed **by use of** very high doses of this drug in combination with other drugs.

- □ development [48,509]：名 発生，発症，開発
- □ resistance [20,333]：名 抵抗性
- □ delay [10,057]：動 遅らせる
- □ by use of ～ [1,116]：～の使用によって
- □ dose [29,485]：名 用量，投与量
- □ in combination with ～ [2,878]：～と組合わせて

訳 抵抗性の発生は，他の薬剤との併用とこの薬剤の非常に高い用量の使用によって遅れるかもしれない

本書に掲載の文例一覧

Ⅰ. 動詞の使い方

● 1 他動詞＋目的語の文例

A. 他動詞＋名詞

1 To further investigate this possibility, we assessed the roles of two other proteins important for cell elongation.

2 The bovine cell grafts in pigs caused an increase in antibodies detected in serum samples.

3 The results demonstrated the presence of a maternal autoimmune disease.

4 We sought to determine the effect of G protein-coupled receptor kinase on bone formation in vivo.

5 These results suggest that this is a valuable model for elucidating the mechanism of prion conversion.

6 Mouse models can be useful in evaluating the effect of a human candidate gene mutation on an intermediate trait.

7 The aim of the current study was to examine the role of loud noise in acoustic neuroma etiology.

8 These genes are known to have a role in an mRNA degradation pathway called RNA interference.

9 These animal antiapoptotic genes have the potential to generate effective disease resistance in economically important crops.

10 These mutations increased the rate of recombination between DNA sequences that had a high degree of sequence homology.

11 These observations indicate the presence of allergen-specific patches consisting of an unusually high proportion of surface-exposed hydrophobic residues.

12 In vitro, Chlamydia pneumoniae can induce the expression of varied molecules in infected human endothelial cells.

13 We systematically investigated the effects of these remarkable new agents in children and adolescents.

14 Ultrasonographic examination of the liver provides evidence of specific patterns of fibrosis.

15 These findings provide a mechanism to explain patterns of gene expression in breast cancer, colon cancer, and malignant melanoma.

16 The results indicated that vaginal epithelial application of these synthetic oligonucleotides reduced the number of animals that developed signs of genital herpes.

17 These orphan nuclear receptors have been proposed to regulate the expression of detoxifying enzymes and transporters.

18 We previously reported the identification of a novel human protein that shuttles between the nucleus and cytoplasm.

19 These mutants retained the ability to cause lethal infections, and thus enabled us to analyze the virulence factors in a surrogate animal model.

20 Sequencing analysis revealed the presence of several known genes.

21 These results suggest a role for these pathways in the context of normal cell proliferation.

22 These data support a model in which macrophage phagocytosis is coordinately regulated by both phospholipases.

23 This study was conducted to test the hypothesis that inefficient metabolism in blood vessels promotes vascular disease.

24 To understand the mechanisms of the mosaic formation, it is of paramount importance to identify evasion mechanisms used by virulent microorganisms.

B. 他動詞＋that節

25 The results showed that increases in neuronal spike rate were accompanied by immediate decreases in tissue oxygenation.

26 The result indicates that PS1 mutations modulate intracellular calcium signaling pathways.

27 The findings imply that chemokine networks serve important functions at the maternal-fetal interface.

28 Data presented here demonstrate that biofilm formation was evident in all specimens.

29 Physicians should note that many months usually pass between the diagnosis of cancer and the occurrence of complications.

30 We conclude that patients with some endogenous insulin secretory capacity do not depend on insulin for immediate survival.

31 Authorities estimate that at least 3 million children suffer from lead poisoning.

32 Pathologic studies reveal that the most likely primary site of origin includes the colon and kidneys.

33 No epidemiologic studies have yet confirmed that women with migraine headaches have a greater risk of stroke.

34 We report that the majority of cells in the hypothalamic circadian pacemaker are lost in patients with senile dementia.

35 We propose that these genes are members of MAP kinase signaling pathways.

36 We hypothesized that the mechanism of genomic integration may be similar to transposition.

37 We speculate that such dysfunction may be relevant to the mutation.

C. 他動詞＋ whether 節

38 The aim of this study was to investigate whether T cells infected with HIV are more susceptible to Fas-induced death.

39 This study examined whether the adverse effects of prenatal exposure to tobacco on lung development are limited to the last weeks of gestation.

40 In this study, we asked whether distinct immunochemical reactions might occur after xenotransplantation of the lung.

41 The studies reported here assess whether a defective oxidative defense may contribute to Down's syndrome.

42 Further studies are required to evaluate whether cellular uptake of DNA is a significant barrier to efficient transfection in vivo.

43 Few animal studies have been carried out to address whether adolescent nicotine exposure exerts unique or lasting effects on brain structure or function.

D. 他動詞＋ to *do*

44 We attempted to identify risk factors for breast cancer.

●2　他動詞（過去分詞）＋前置詞／that節の文例

A. 他動詞（過去分詞）＋前置詞＋名詞

45 Much has been learned about the underlying mechanisms for adverse effects of chemotherapy.

46 The suprachiasmatic nucleus is regarded as the main mammalian circadian pacemaker.

47 The prediction of protein structure is viewed as a great challenge for scientists.

48 This study is aimed at identifying the signaling pathways involved in these events.

49 Industrial aerosol exposure leads to headache, drowsiness, nausea, and vomiting, followed by a latent period of 1 to 5 days before the onset of additional symptoms.

50 Molecular interactions are required for normal signal transduction across the lipid bilayer.

51 The patient should be observed for evidence of proximal colonic bleeding.

52 This virus-like particle was discovered in an infectious stool filtrate derived from an outbreak of gastroenteritis in a preschool.

53 Several potential binding sites can be inferred from functional mapping.

54 Several growth factors implicated in the stimulation of fibroblasts are capable of enhancing the production and secretion of extracellular matrix proteins.

55 The lesions of lymphoma are often found to be located in the white matter adjacent to the ventricles.

56 This protein product expressed on the surface of tumor cells structurally appears to be a tyrosine kinase receptor analogous to the epidermal growth factor receptor.

57 Specialized tests should be performed on an appropriately prepared tumor biopsy to diagnose responsive endometrial cancer.

58 Five-year survival of interferon-treated patients was compared with that of those treated with chemotherapy.

59 These results suggest that fat intake is correlated with the risk for colorectal cancer.

60 Animal-associated opportunistic infections were reported to be found among persons infected with the human immunodeficiency virus.

B. 他動詞（過去分詞）＋to *do*

61 The numbers are expected to increase by 20% in the next decade.

62 The gene is predicted to encode an integral membrane protein.

63 These responses are presumed to be modulated by protein-protein interactions.

64 Aβ accumulation has been postulated to contribute to the pathogenesis of Alzheimer's disease.

65 Cyclin D1 is thought to be a key regulator involved in cell cycle progression.

66 The vitamin D receptor is believed to mediate this activity.

67 Acetylation of histones is considered to be a critical step in transcriptional regulation.

68 Wnt signaling is known to be involved in early steps of neural crest development.

69 This novel approach can be used to assess quantitatively the effects of therapeutic interventions for treating liver failure.

70 The present study was designed to evaluate this hypothesis.

71 This study was undertaken to establish a rat model of chronic rejection.

C. itを形式主語とするthat節

72 It is widely accepted that tumors are monoclonal in origin.

73 It is generally assumed that protein kinase C is the sole receptor for phorbol esters in this system.

● 3 自動詞＋前置詞の文例

A. 自動詞＋前置詞

74 There are currently no vaccines to protect against a number of serious viral diseases such as SARS and AIDS.

75 Mites are known to serve as a vector and reservoir of the etiologic agent.

76 Environmental agents producing point mutations may act as chromosome-breaking agents.

77 Structural abnormalities resulting from errors in embryogenesis or the fetal period are called congenital anomalies.

78 Malignant lung tumors commonly arise from the respiratory epithelium.

79 Bacteria differ from fungi in the lack of ability to reproduce sexually or asexually.

80 Headache may originate from damage to pain-sensitive pathways of the peripheral or central nervous system.

81 The combined treatment resulted in improved survival.

82 The BRCA genes are suspected to participate in the development of sporadic breast cancer.

83 The objective of this study was to determine whether differences exist in the trabecular bone structure of the femur and tibia.

84 Embryonic vessels differentiate into arteries and veins.

85 G protein α subunits consist of two domains: a GTPase domain and a helical domain.

86 Depending on the dose administered, cystic fibrosis mice exhibited significantly higher mortality rates, compared to wild-type mice.

87 This review will focus on tissue engineering as a promising approach for cartilage regeneration and repair.

88 The test should not rely on the patient's ability to monitor or accurately record blood glucose levels.

89 The downstream events leading to retinal damage are poorly understood.

90 Despite decades of investigation, how insulin binds to its receptor remains largely unknown.

91 Human memory T cells respond to microvascular endothelial cells and can injure allografts in vivo without priming.

92 A number of fertilized oocytes failed to develop beyond the two-cell stage.

93 An intracellular protein known as Calmodulin interacts with a number of proteins involved in signal transduction.

94 The suppressor P25 may interfere with either assembly or function of the effector complexes of RNA silencing.

95 We sought to prepare a T cell population capable of reacting with most adenoviruses that cause disease in immunocompromised patients.

II. 副詞の使い方

1 副詞＋過去分詞の文例

96 These mutations were distributed in highly conserved residues and were absent in 300 control chromosomes from an ethnically similar population.

97 The prevalence of diarrheal diseases should be significantly reduced in developing countries through improvements in hygiene to limit fecal-oral spread of enteric pathogens.

98 A serine/threonine kinase expressed in the heart has been previously reported to regulate sodium channels.

99 It remains unresolved whether arthropods are more closely related to nematodes or to deuterostomes.

100 In some countries, a live anthrax vaccine has been widely used for prophylaxis against anthrax in both humans and animals.

101 The aim of this study was to identify differentially expressed genes in rat fetal liver epithelial stem cells during their proliferation, lineage commitment, and differentiation.

102 The levels of Sp1 protein and mRNA were markedly reduced at postnatal day 15.

103 In this study, we used estrogen-dependent human breast cancer cells stably transfected with the aromatase gene.

104 Secretion of newly synthesized proteins across the mammalian rough endoplasmic reticulum is known to be supported by two membrane proteins.

105 These results raise the possibility that a positively charged loop 2 is important to maintain processivity near physiologic ionic strength.

106 In lean mice, food intake was almost completely blocked by the fatty acid synthase inhibitor, C75.

107 The border cells in the ovary undergo a well-defined and developmentally regulated cell migration.

108 Upstream stimulatory factor seems to be necessary for full promoter activity in transiently transfected cells.

109 These survival and apoptotic signals are tightly regulated by a large number of molecules.

110 Improved recombinant DNA technology is expected to lead to development of genetically engineered inhibitors, such as dominant-negative mutants.

111 Replacement therapy using partially purified human enzyme has proved to be biochemically effective in short-term pilot trials.

112 Cognitive development is severely impaired in individuals with Williams syndrome.

113 The mechanisms that cause these translocations are not fully understood.

114 A nucleotide containing a phosphate group complexed with magnesium ion appears to be absolutely required for binding.

115 The role of this regulatory factor in vertebrate myogenesis has been extensively studied.

116 Patients with defects in cellular immune competence are more likely to remain chronically infected rather than to clear the virus.

117 We observed rapidly induced changes in a cortically mediated perception in human subjects.

118 We performed a comparative analysis of all protein-antibody complexes for which structures have been experimentally determined.

119 The mechanisms responsible for bone resorption inhibition have not been clearly defined.

● 2 副詞＋形容詞の文例

120 There were no statistically significant differences in demographics between the experimental group and the control group.

121 Transgenic mice expressing constitutively active protein kinase C in the colon were found to be highly susceptible to carcinogen-induced colon carcinogenesis.

122 We sought to evaluate physiologically relevant interactions between ligands for this versatile transport protein.

123 Relatively little is known about the peripheral mechanisms that excitatory amino acid receptors may regulate when activated.

124 These novel approaches may serve as potentially important therapeutic interventions in the treatment of advanced colorectal cancer.

125 Smoking is a much larger risk factor for cardiovascular disease mortality than fine particulate matter air pollution.

126 These two proteins were detected in approximately equal amounts in the cell.

127 These eight genes were found to be mutated in a mutually exclusive manner.

128 The inner ear defects in these mice are remarkably similar to those seen in the corresponding human condition.

129 Distinctly different patterns of gene expression were observed at different levels of osteoprogenitor maturation.

130 In antibody testing, females had a slightly higher incidence of positivity than males.

131 The activated proteases are primarily responsible for the destruction of the bacteria.

● 3 副詞＋前置詞の文例

132 β-adrenergic blockers lower arterial pressure directly by reducing systemic vascular resistance.

133 Human herpesvirus 6 is an immunosuppressive and neurotropic virus that may be transmitted by saliva and possibly by genital secretions.

134 Suramin actively treats prostate cancer, presumably by blocking the action of growth factors such as FGF.

135 These results indicate that the toxic effects of glutamate receptor overstimulation can be explained solely by calcium influx.

136 The neuropeptide galanin gene expression in the cholinergic basal forebrain may be negatively regulated, perhaps by factors concomitant with puberty.

137 ABCC6 is expressed primarily in the liver and to a lesser degree in the kidney.

138 This orphan receptor tyrosine kinase is expressed almost exclusively in endothelial cells.

139 Lymphangioleiomyomatosis is the result of aberrant smooth muscle proliferation and occurs in a sporadic form predominantly in premenopausal women.

140 Elemental mercury in blood is excreted mainly in the urine and feces.

141 This membrane-active antifungal agent was found to bind specifically to amyloid fibrils.

142 Each of the recombinant proteins was found to bind preferentially to a specific fragment of minicircle DNA.

143 These proteins were engineered to respond rapidly and selectively to their target.

144 These results, together with previous reports, suggest that the expression of an unidentified DNA polymerase may account for the mutant phenotype.

145 The seroprevalence of the human T-cell leukemia virus type 2 was found to increase linearly with age.

146 These infections are clinically indistinguishable from Mycoplasma pneumoniae infection and may in fact occur simultaneously with mycoplasmal infections.

● 4 文頭の副詞の文例

147 Finally, significant progress was achieved in the development of improved vectors for future gene therapy of human diseases.

148 Interestingly, infected epithelial cells expressed cytokines that augment γ interferon production.

149 Additionally, six residues exerting significant effects on OCT binding were also found within the putative cleft region.

150 Surprisingly, the inhibitor-treated cells remained pluripotent despite the absence of leukemia inhibitory factor.

151 Recently, a new hypothesis has been put forward suggesting that metabolic switching is an intrinsic property of skeletal muscle.

152 Conversely, enzymatic and structural analyses showed that the mutation can affect substrate specificity for phosphotyrosine.

153 Previously, we demonstrated that insulin facilitates muscle protein synthesis in obese Zucker rats.

154 Consequently, it is suggested that fetal catecholamines play an important role not only in altering fetal metabolism but also in regulating fetal growth.

155 Subsequently, retinoic acid was detected at relatively high levels in the central nervous system of adult rats.

III. 形容詞の使い方

● 形容詞＋前置詞の文例

156 Lead is thought to be responsible for an increased incidence of spontaneous abortion because it readily crosses the placenta.

157 Trace elements such as iron, zinc, copper, chromium, selenium, iodine, fluorine, and manganese have been identified as essential for health in humans.

158 The residue was found to be important for nuclear localization and DNA binding.

159 The aim of this study was to determine whether stem cells are necessary for epidermal renewal.

160 Monoclonal antibodies specific for vaccinia virus have been commercially available.

161 The degradation of local basement membrane is not sufficient for distant metastasis of malignant cells.

162 This model may be useful for studying the role of posttranslational modification.

163 Angiogenesis is crucial for tumor growth.

164 An array of diagnostic tests is now available for evaluation of patients with evident or suspected cardiovascular disease.

165 Seven of 95 specimens tested positive for HIV antibodies.

166 Severe developmental abnormalities have been reported in mice homozygous for this enzyme deficiency.

167 This secretion signal seems to be distinct from the N-terminal signal peptide.

168 Survival was significantly different from what was expected for an age- and sex-matched population.

169 Mice lacking one of these receptors are indistinguishable from their wild-type littermates.

170 Major congenital anomalies were confirmed in 3 to 4% of all newborn infants, independent of ethnic group.

171 The therapeutic effects on proteinuria, renal pathologic features, and survival rate were found to be dependent on the drug dose.

172 Diagnostic tests and other clinical evaluations help to identify tumors that are sensitive to this particular drug.

173 The two groups of strains were extremely resistant to the bactericidal activity of serum.

174 The pattern of expression of proapoptotic genes was nearly identical to the pattern of apoptotic genes induced by tumor necrosis factor α.

175 The results obtained by this rapid, reproducible, and noninvasive method, computed tomography, were comparable to those obtained by histology.

176 Randomized double-blind trials should be carried out to detect common adverse reactions attributable to this vaccine.

177 These findings are consistent with the hypothesis that the functional maturation of the primate visual brain proceeds in a hierarchical manner.

178 These in vitro findings were compatible with in vivo observations of synapse growth and elimination in living tadpoles of Xenopus laevis.

179 The activation of mitogen-activated protein kinase was coincident with the age-dependent increase in amyloid deposition and loss of synaptophysin.

180 Concomitant with the reduction in spheroid number, the mutant mice showed increased viability and an improved behavioral phenotype.

IV. 名詞の使い方

● 1 名詞＋前置詞の文例

181 The present study provides biochemical information about the substrate specificity of the enzyme.

182 Per1 mRNA levels peaked at 2 hours after stimulation.

183 Protection against apoptotic insults may arise through the inhibition of a proapoptotic protein.

184 Interleukin-4 cytotoxin exhibited remarkable antitumor activity against ovarian tumors in immunodeficient animals.

185 Defense against host-derived reactive oxygen species has previously been reported as critical for intracellular replication.

186 The mean age at onset of symptoms was 6.7 years and at diagnosis was 7.3 years.

187 A complex relationship between food and health in our society is intrinsically linked to the obesity epidemic.

188 To date, no studies have evaluated associations between human leukocyte antigen class II markers and posttreatment chronic Lyme disease.

189 There were no identifiable differences between adults and children with respect to the sensitivity and specificity of serologic tests for celiac disease.

190 There was a strong correlation between changes in the number of tumor cells and clinical status.

191 This discovery suggests a possible link between aging and stem cell dysfunction.

192 Investigations of leprosy provide new insights into the mechanisms by which the innate immune system contributes to host defense against infection.

193 Insulated from the external environment, the pineal gland responds to discontinuous activation by the circadian clock.

194 In order to gain insights into acute rejection, we investigated MHC class II expression by the allograft.

195 We sought to evaluate the genetic contribution to sensitivity to cell growth inhibition by anticancer agents.

196 These data collectively offer compelling evidence for the transmembrane segment providing a new model of this protein.

197 We have developed a quantitative method for determining protein interactions in cells.

198 Collectively, these results may provide a cellular and biochemical basis for heart failure.

199 Dextrose infusion leads to a decrease in the requirement for amino acid oxidation as an energy source.

200 Zinc is bound to metallothioneins synthesized in the liver and has a strong affinity for red cells and plasma proteins.

201 This enzyme appears to play a role in controlling cell cycles and can be an important target for new chemotherapy agents.

202 These results highlight the need for further studies of the effect of isoflavones on bone.

203 These findings have implications for the design of future clinical trials in patients with prostate cancer.

204 A strong candidate gene for Alzheimer's disease was identified in a region at chromosome 12.

205 The authors present a novel strategy for generating and analyzing comprehensive genetic-interaction maps.

206 Recent advances in understanding the molecular mechanisms of the differentiation process will hopefully lead to more effective therapies for lung cancer.

207 This method has been used as a tool for localizing the effect of visual stimuli.

208 Fission yeast can serve as an excellent model system for genetic analysis of cell-polarity determination.

209 This method has great potential for identifying the pathways that are altered in response to the mutant protein.

210 The aim of this study was to characterize pathways for the uptake and intraneuronal trafficking of protease-resistant prion protein.

211 After adjustment for age and sex, there was no significant association between race and the risk of death from heart disease.

212 These data provide additional support for the hypothesis that this technique could elucidate the tissue of origin of metastatic lesions.

213 Together, these findings provide a plausible explanation for the severity of the autoimmune diseases in the mutant mice.

214 Rising incidence of tuberculosis has prompted the search for alternative biomarkers using newly developed postgenomic approaches.

215 This negative-strand RNA virus has inherent specificity for replication in tumor cells due to their attenuated antiviral responses.

216 Using data from an observational study of 300 healthy adults, we analyzed the association of body mass index with dietary intake of carbohydrates.

217 In these mice, serotonin levels remained unchanged, and dopamine uptake and release from nerve terminals were normal.

218 Fecal samples from healthy children who received oral poliovirus vaccine were found to contain variants of Sabin vaccine viruses.

219 We assessed differences in serum hyaluronan levels between the two groups, adjusting for ethnicity, sex, age, and body mass index.

220 A significant reduction in ejection fraction, as assessed by repeated echocardiography, was also observed.

221 Mobilization of endothelial progenitor cells with cytokines was found to enhance bone marrow cell incorporation into ischemic myocardium.

222 Data on the association between androgen receptor polymorphisms and ovarian cancer are inconclusive.

223 The identification of numerous microtubule-associated proteins may have a profound impact on the study and treatment of human genetic disease.

224 There have been few studies on the mechanisms that underlie the function of this novel isoform encoded by an alternatively spliced transcript.

225 This study provides new information on the molecular basis of resistance and the evolution of resistance.

226 The major histocompatibility complex polymorphism has critical influence on the properties of the selected cytotoxic T lymphocyte repertoire.

227 The enzyme displayed a strong dependence on the ionic strength of the buffer.

228 This transient loss of tetramer binding is associated with reduced signaling through the T-cell receptor.

229 Chronic bronchitis is associated with impaired resistance to bronchial infection.

230 Accumulating data emphasize the role of genetic factors as a cause of increased susceptibility to adverse drug responses.

231 Type 2 diabetes occurring spontaneously in rhesus monkeys shows a striking similarity to human diabetes in clinical features.

232 RNA-binding motif is known to have homology to RNA-binding proteins, but its function remains largely unknown.

233 Treatment with this monoclonal antibody alone had relatively little effect on survivin and apoptosis.

234 When red cells are damaged, hemoglobin is liberated into the circulation and forms a complex with circulating haptoglobin.

● 2 名詞＋that（同格のthat）の文例

235 This suggestion is based on the observation that at least 20 % of sporadic melanomas arise in association with atypical cells.

236 These findings support the idea that high iodine intake can induce autoimmune thyroiditis in genetically predisposed animals.

237 Of some concern is the notion that obstructive sleep apnea may result in sudden death during sleep.

238 These results are compatible with the view that metabolic environment per se causes complications independent of genetic factors.

V．つなぎの表現

1　逆説の文例

A．副詞／副詞的熟語

239 However, the underlying mechanisms remain poorly understood.

240 Nevertheless, some of the patients retained sufficient cognitive and verbal activity to perform mental status examinations at a normal level.

241 Nonetheless, infection and rejection are the two major problems.

242 Instead, the interplay of various intracellular signaling pathways probably account for the functional synergy.

243 Alternatively, computer-based clinical support systems should be introduced that make it possible for physicians to utilize optimal antibiotic choices.

244 On the contrary, no significant changes in cytokine levels were found between the patients and healthy volunteers.

245 In contrast, anti-$\gamma\delta$ antibody potently inhibited proliferation of $\gamma\delta$ T cells.

246 By contrast, there are no currently available vaccines for the bacterium responsible for syphilis.

247 On the other hand, some forms of cancer therapy have been reported to be very effective and well tolerated even with advanced age.

B．副詞節を導く接続詞

248 Although all communication between neurons is known to occur through synapses, little is known about the mechanisms inducing their formation.

249 Even though the process of transdifferentiation has been well studied and established in amphibian systems, whether mammalian cells possess the same potential remains unclear.

250 While withdrawal symptoms are present in late stages of drug dependence, physicians should also be alert to the early behavioral signs of impaired social functioning.

251 Seventy-five percent of the examined human neurofibromas expressed progesterone receptor, whereas only 5% expressed estrogen receptor.

C. 副詞句を導く熟語／接続詞

252 In contrast to other translation initiation factors, these two factors colocalize to a specific cytoplasmic locus.

253 The disagreement of the value with the results of other observers is explained in terms of individual metabolic properties as opposed to those of the larger population.

254 Contrary to the widely held assumption, rapid eye movement (REM) and non-REM sleep are not mutually exclusive states.

255 Despite extensive data regarding the in vitro and in vivo activities of the drugs, early studies did not recognize that the high concentrations of these drugs used in vitro were unattainable in vivo.

256 In spite of extensive research on Parkinson's disease and Alzheimer's disease, preventive or long-term effective treatment strategies have not yet been proposed.

257 Unlike renal disease, no useful laboratory test is available to make any general recommendations for drug dose adjustments in liver disease.

● 2 肯定の文例

A. 副詞／副詞的熟語

258 Therefore, the physician should always take into consideration the possibility of an adverse drug response in the differential diagnosis.

259 Hence, the most intense monitoring is needed in the first 6 to 8 weeks after transplantation.

260 Accordingly, only a few antivirals for respiratory viral illnesses are approved for use, each with very limited indications.

261 Thus, the increased risk of dying of pneumonia is probably related to the decreased ability of the aged immune system to combat the pathogens.

262 Those infected adults were selected and then screened for drug-selected resistance mutations and phylogenetic subtype.

263 The aim of the drug therapy was to induce vascular smooth muscle relaxation, thereby relieving spasm, raising resting blood flow, and limiting the degree of ischemia during attacks.

264 Furthermore, our data suggest that genital epithelial cells may provide a barrier to HIV infection.

265 Further, murine embryonic fibroblasts obtained from these mice did not die in response to oxygen deprivation.

266 Moreover, the combination therapy was more effective than temozolomide treatment alone.

267 In addition, immunosuppressive therapy in animal trials has been reported to increase early mortality.

268 Indeed, nonmammalian nervous systems provide ideal platforms for the study of fundamental problems in neuroscience.

269 In fact, these drugs may be harmful under certain circumstances.

270 Similarly, ineffective helper T cell activation inhibits the immune response to blood group antigens in ABO-mismatched allograft recipients.

271 Likewise, the physical examination needs to be complete, with more emphasis placed on carefully measuring the blood pressure.

B. 副詞節を導く接続詞／熟語

272 Because ischemia can evoke the generation of reactive oxygen species, we explored the effect of oxidative stress on membrane phospholipid.

273 Since the cause of this disease is unknown and no cure is available, the physician must distinguish as early as possible the attendant risks for each patient.

274 The doctor must work to develop a partnership with the patient so that he can identify and use diagnostic tests and treatments that are acceptable to the patient.

C. 副詞句を導く熟語

275 In agreement with these findings, knockdown of Wnt signaling resulted in increased cell proliferation.

276 In accordance with these data, plasma corticosterone levels were decreased.

277 Sleepiness peak typically occurs during the latter half of the night, coincident with maximal sleep drive within the brain.

278 According to this mechanism, the enzyme undergoes multiple conformation changes during the catalytic cycle.

279 In addition to abnormalities of muscle protein metabolism, heritable factors may contribute to susceptibility to external triggers for cardiac dysfunction.

280 Because of the biotechnology revolution, many novel vaccines will be developed and become available in the future.

281 Due to the lack of adequate animal models, the function of these molecules in disease pathogenesis remains poorly understood.

282 Imaging technology has advanced the accuracy of diagnosis owing to increased sensitivity.

3 まとめの文例

A. 副詞／副詞的熟語

283 Taken together, these results demonstrate that alterations of mitochondria potentially contribute to aging and age-related disease in the nervous system.

284 In summary, a number of critical questions remain unanswered concerning the interactions between alcohol and hepatitis C.

285 In conclusion, progenitor cell accumulation in damaged livers has been clearly shown to be modulated by stress-related sympathetic activity.

4 条件の文例

A. 条件

286 If any of these reaction steps are rate-limiting to turnover, the substrate for the rate-limiting reaction should accumulate.

287 The physician should not rule out the possibility of an adverse drug reaction in the differential diagnosis even if none has been reported previously for the particular drug.

288 Once the genetic model has been established, the facts based on it can be clearly explained to the patient and family.

289 Given limited advances in renal cell carcinoma, a thorough understanding and testing of rationally targeted agents is needed.

VI. その他の表現

1 比較の表現の文例

A. ～than

290 More than 15% of those older than 65 years are reported to be suffering from severe depression.

291 Age rather than menopausal status is a significant independent predictor of vascular events in women with systemic lupus erythematosus.

292 Physical abuse is a leading cause of serious head injury in children less than three years of age.

293 These signal transduction pathways are required for greater than 90% of the synergistic response.

294 Practicing physicians should be aware that the rate of cardiac complications after smallpox vaccination has been higher than expected.

295 A single intraperitoneal injection of this iron regulatory hormone resulted in iron levels 80% lower than in control mice.

296 The aim of this study was to examine the influence of age at menopause on specific causes of death other than coronary heart disease.

B. ～-fold, %, times

297 This novel fluorescence assay provided a 20-fold increase in sensitivity with a broad dynamic range.

298 A subset of these genes exhibited an approximately 2-fold decrease in retinal gene expression.

299 A 45% reduction in accumulation of these cells was observed in an acute lung injury model.

300 Although the men and women showed similar insulin responses to feeding, leptin concentrations increased 3-fold only in women.

301 While low doses dramatically enhanced immune responses, 10-fold higher doses did not augment responses.

302 Although azide and formate exhibited different potencies as scavengers, the former was almost 2-fold more potent than the latter.

303 Females are two to three times more likely to be affected by rheumatoid arthritis than males.

304 The AIDS rates for Hispanic men were reported to be three times higher than for white men in the US.

C．compared, comparison, relative

305 Compared with wild-type mice, these mice displayed markedly exacerbated disease progression and pathology.

306 Screening tests for lung cancer with annual chest x-ray films and sputum cytology resulted in no improvement in overall mortality in comparison with control subjects.

307 Rodent models of obesity showed significant increases in food intake and weight gain relative to controls.

D．to a lesser ～／… degree of ～

308 Histologically, iron is found in increased amounts in organs such as the liver, heart, and pancreas and to a lesser extent in the endocrine glands.

309 Medications with a low therapeutic index known to have a high risk of drug interactions should raise a high degree of suspicion of adverse drug interactions.

2 asを用いた表現の文例

A. as＋過去分詞

310 As compared with placebo, both classes of drugs demonstrated efficacy in the treatment of rheumatoid arthritis.

311 DEC1 physically interacted with Bmal1, as determined by coimmunoprecipitation.

312 These osteoblasts were shown to produce nerve growth factor, as measured by ELISA.

313 As expected, these young mice showed markedly elevated apoptosis compared with wild-type controls.

314 Environmental factors are considered important in the etiology of gastric cancer, as evidenced by Japanese descendents in Hawaii showing a decreased incidence of gastric cancer.

315 These symptoms are accompanied by increasing evidence of damage in the basal ganglia as shown by magnetic resonance imaging.

316 One of the symptoms of dementia is impairment in abstract thinking, as indicated by difficulty in defining words and concepts previously known.

317 This effect is gene-specific, as demonstrated by the fact that G3PDH expression is not affected.

318 These three binding sequences accounted for about 65% of the transcriptional effect, as judged by transient transfection assays.

319 Here, we report studies detailing the cross-sectional imaging of bowel obstruction as opposed to the more traditional imaging with barium radiography.

320 Besides ultraviolet radiation from sun exposure, there appear to be other causal agents in the development of cutaneous melanoma as observed in epithelial carcinomas.

321 Unilateral removal of the avian cochlea caused a drastic reduction in the expression of these proteins in the NM neurons as detected by immunocytochemistry.

322 Mutants in this particular gene exhibited asynchronous overreplication during normal growth, as revealed by flow cytometry.

323 The mean age of onset of puberty for girls as defined by breast budding is a little earlier than that for boys as defined by enlargement of testes.

324 As described in the accompanying article, our study represents the first evidence of a super-integron in a non-pathogenic bacterium.

325 As predicted by the model, we found genetic evidence that these cells are sufficient to alter neural activity in regions involved in autonomic and neuroendocrine control.

326 These results underscore the role of this pathway in the biology of plasma cell growth as reflected by its influence on survival.

B. as ＋名詞

327 Family members taking care of a dementia patient sometimes develop clinically significant symptoms of depression as a result of the stress associated with prolonged caregiving.

328 Many African countries are urged to complete measles supplemental immunization activities in children aged 9 months to 14 years as part of a comprehensive measles-control strategy.

329 Thrombosis of the renal arteries and segmental branches may arise as a consequence of intrinsic pathology of the renal arteries.

330 This long single-stranded tail is expected to serve as a substrate for telomerase.

331 A wealth of data for exercise tolerance testing can be used as a means of identifying asymptomatic patients at high risk for coronary heart disease.

332 This antigen is known to serve as a marker of HBV infection.

333 Sixty-seven subjects without prominent visual symptoms were selected as controls.

334 The kidney is a major excretory organ and serves as a target for many hormones.

335 Use of the cDNA as a probe enabled us to determine the onset, relative levels, and locations of gene expression in various adult tissues.

336 The method introduced here is practical and available to the target population as a whole.

337 We reviewed the evidence of apoptosis as a mechanism for this tumor response in p53 mutant breast cancer.

338 The role of genetic factors as a source of interindividual variation in drug response seems to be well established.

339 This gene has been reported to act as an inhibitor of neuronal cell proliferation in the secondary neural tube.

340 This homolog of the protein has been implicated as a component of the viral budding machinery.

341 As a first step, we have embarked on the characterization of this enzyme in zebrafish.

342 As an example, we show how contemporary parenteral products with microbial contaminants can be considered safe under current pharmacopoeia tests, but provoke adverse clinical effects.

343 These abnormalities found in patients with renal failure have been used as an index of the adequacy of hemodialysis.

344 Another potent inhibitor of platelet function is being used as an alternative to aspirin in patients with cerebrovascular disease.

345 Evidence-based medicine requires scientific methods as the basis for understanding and treating disease.

C. as ～ as

346 Frequent injuries caused by falls are seen in sedative abusers as well as alcoholics.

347 This inhibition was dose-dependent, and was observed as early as 3 hours after stimulation.

348 The spleens of these mice are as much as 30% larger than those of wild-type littermates.

349 The maximum response was observed at a glucose concentration as low as 11 mM.

350 Resistance to this antibiotic in Streptococcus pyogenes was found to be as high as 48% in specific populations in the United States.

351 These patients were followed up longitudinally during and after chemotherapy for as long as 5 years.

352 This new method has made it possible to obtain a stack of axial images and reproduce them in as little as 15 seconds.

353 These data suggest that androgen treatment may be as effective as estrogen replacement in reversing the depression at hippocampal synapses.

354 As many as one in 20 people in the United States have some form of autoimmune disease.

355 Many of these peptides are very small, comprising as few as six residues.

356 The virus replicated as efficiently as wild-type virus in these cells.

D. as ＋前置詞

357 The question arises as to whether these physiologically indistinguishable cells have any special functional correlates.

358 As for gastrointestinal cancer, chemotherapy and radiotherapy have limited efficacy unless the tumor is localized.

359 As with all complex traits, multiple sclerosis is caused by an interplay between unidentified environmental factors and susceptibility genes.

360 As of this writing, the novel genome browser application has been set up on the human, mouse, and rat genomes.

E. asで始まる節

361 As thallium is excreted in the urine, thallium determinations can be made on 24-hour specimens.

362 Particularly difficult times can arise as physicians begin dealing with patients who are becoming old, frail, or cognitively impaired.

VII. 熟語表現

● 熟語表現の文例

363 This mechanism of DNA binding may play an important role in maintaining stable negative regulation of this particular gene expression in the absence of extracellular stimulation.

364 After adjusting for body mass index, age, and sex, there was no significant difference in the serum cholesterol concentrations.

365 Our results obtained with the molecular dynamics simulation technique are in good agreement with the data available from experiments.

366 In an attempt to circumvent this adverse effect, we evaluated the efficacy of aerosol and oral vaccinations.

367 On the basis of these findings, we propose that tyrosine phosphorylation is not involved in the signal transduction pathway leading to cell growth suppression.

368 Blood pressure and peripheral vascular resistance fell at the beginning of pregnancy.

369 In the case of rare hereditary disease, a careful family history is indispensable in assessing and understanding the disease.

370 Congenital forms of ACTH deficiency usually occur in combination with the loss of other pituitary hormones.

371 The estrogen receptor, in common with other nuclear hormone receptors, contains two transcription activation functions.

372 We analyzed the crystal structures of other bone morphogenetic proteins in complex with their receptors.

373 Without exception, these compounds were all inactive at a concentration of 10 μM.

374 Further studies are needed to assess the utility of highly selective inhibitors of glycogen synthase kinase-3 for the modification of insulin action under conditions of insulin resistance.

375 Under the same conditions, cooperativity was much greater for the pancreatic form of the enzyme than for the liver form.

376 Iritis is known to occur frequently in conjunction with intestinal inflammation.

377 The mitral and aortic valves are normally in contact with each other.

378 We discuss our findings in the context of mechanisms that may underlie tumorigenesis.

379 We have established physiological roles and sites of action of the different topoisomerases within the context of the bacterial cell cycle.

380 Each of these transgenes is under the control of individual poxvirus promoters.

381 Response to this drug varies considerably among patients during the course of therapy.

382 Here, we present evidence that this family of transposable elements has been significantly amplified over the course of evolution.

383 This anticancer chemotherapy drug was administered at a dose of 1 mg/kg per day for 3 days.

384 In an effort to characterize mechanisms common to these processes, we describe the earliest stages of offspring formation in this prokaryote.

385 At the end of the experiment, the dystrophic rats were perfused and processed for histologic assessment of photoreceptor survival.

386 To this end, we have developed a novel method for the differential enhancement of probe sequence concentration by subtractive hybridization.

387 These data contribute to several lines of evidence indicating that genetically distinct subtypes of motor neurons are specified during development.

388 For example, two recently published reviews investigated the evidence on racial/ethnic differences in medical care.

389 The estimated annual hospital cost is in excess of $7 billion.

390 This may help to preserve the function of the retinal pigment epithelium in the face of aging and disease.

391 These findings constitute strong evidence in favor of the beneficial effects of glucosamine.

392 Ischemic injury to the kidney is characterized in part by tubular cell death in the form of necrosis or apoptosis.

393 The tetraploid cells gave rise to malignant mammary epithelial cancers when transplanted subcutaneously into nude mice.

394 For instance, in some forms of leukemia, younger patients respond to therapy better than older patients.

395 To our knowledge, these results provide the first evidence that haplotypes in the growth hormone secretagogue receptor gene are involved in the pathogenesis of human obesity.

396 In light of recent studies, it is now clear that mutations outside the kinase domain can lead to enhanced autophosphorylation of the kinase.

397 These findings shed light on the molecular function and evolution of clock genes in vertebrates.

398 Physicians should think of anemia in a manner similar to the way in which they regard symptoms like diarrhea as an indicator or a manifestation of an underlying disease.

399 By means of a questionnaire, we identified women at high risk for chlamydial infection.

400 The influenza vaccine was successful in protecting against pneumonia and death on the order of 60 to 80%.

401 Rates of cervical cancer have declined at least in part due to cytologic screening for cervical pathology.

402 A number of alternative antibiotics can be used in place of penicillin.

403 Many clinical decisions require careful dose adjustment of drugs in the presence of renal dysfunction.

404 These modifications of cellular gene expression probably occur during the process of liver cell injury and repair.

405 Increasing evidence demonstrates that risks of lung cancer rise in proportion to both duration of smoking and amount smoked per day.

406 Shortly after the infection, the microorganisms were found in the submucosa in close proximity to local immune cells and blood vessels.

407 Some victims of accidents or work injuries may exaggerate their pain for the purpose of compensation or for psychological reasons.

408 Purulent sputum raises the question of chronic bronchitis, bronchiectasis, pneumonia, or lung abscess.

409 The size of the treated population has been expanding at a rate of 5% per year.

410 With regard to adult stem cells, their use in research and therapy is not controversial because the production of adult stem cells does not require the destruction of an embryo.

411 In support of this hypothesis, we demonstrate that both of these phenotypes are strain dependent.

412 The gene rearrangements took place at various stages of the differentiation pathway.

413 These data show for the first time that exercise performed immediately following immobilization induces profound changes in the expression of a number of genes in favor of the restoration of muscle mass.

414 In view of their potential for malignancy, it is generally recommended that polyps of any size causing significant symptoms be removed.

415 The development of resistance may be delayed by use of very high doses of this drug in combination with other drugs.

索　引

記号

□ ～-associated ―― 53
□ ～-fold decrease ―― 184
□ ～-fold higher ―― 185
□ ～-fold increase ―― 183
□ ～-fold more potent ―― 186
□ …hours after～ ―― 120
□ ～-induced ―― 40, 85
□ ～per day ―― 231
□ ～% reduction ―― 184
□ ～ times higher ―― 186
□ ～ times more… ―― 186
□ ～ times more likely to … ―― 186
□ ～-treated ―― 52, 101

欧字

A

□ aberrant ―― 94
□ ability ―― 63, 67, 162
□ ability to～ ―― 63
□ abnormality ―― 62, 109, 172, 209
□ abscess ―― 242
□ absence ―― 101
□ absent ―― 73
□ absolutely ―― 81
□ absolutely required ―― 81
□ abstract thinking ―― 194
□ abuse ―― 180
□ abuser ―― 211
□ accept ―― 59
□ acceptable ―― 169
□ accident ―― 241
□ accompanied by～ ―― 34, 194
□ accompany ―― 33
□ accompanying ―― 198
□ Accordingly, ―― 161
□ according to～ ―― 171
□ account for～ ―― 97, 150, 195
□ accumulate ―― 142, 176
□ accumulation ―― 55, 175, 184
□ accuracy ―― 173
□ accurately ―― 67
□ acetylation ―― 56
□ achieve ―― 99
□ acoustic ―― 24
□ act ―― 62
□ act as～ ―― 62, 207
□ ACTH ―― 226
□ action ―― 92, 228, 229
□ activate ―― 86, 89
□ activation ―― 115, 166, 226
□ activation by～ ―― 125
□ active ―― 95
□ actively ―― 92
□ activity ―― 56, 149, 158, 175, 199, 202
□ activity against～ ―― 121
□ acute ―― 126
□ acute lung injury ―― 184
□ additional ―― 134
□ Additionally, ―― 100
□ address ―― 43
□ address whether～ ―― 43
□ adenovirus ―― 70
□ adequacy ―― 209
□ adequate ―― 173
□ adjacent ―― 51
□ adjacent to～ ―― 51
□ adjust ―― 137
□ adjusting for～ ―― 138
□ adjustment ―― 159, 240
□ adjustment for～ ―― 133
□ administer ―― 66, 231
□ adolescent ―― 27, 43
□ adrenergic blocker ―― 91
□ adult ―― 124
□ advance ―― 131, 173, 177
□ advanced ―― 86
□ advanced age ―― 152
□ adverse ―― 142, 161, 190, 208
□ adverse drug reaction ―― 176
□ adverse effect ―― 41, 46, 224
□ adverse reaction ―― 114
□ aerosol ―― 48, 224
□ a few ―― 161
□ affect ―― 102, 195

□ affected ―― 186
□ affinity ―― 128
□ affinity for～ ―― 128
□ after adjusting for～ ―― 223
□ after adjustment for～ ―― 133
□ after stimulation ―― 212
□ age ―― 122
□ age at～ ―― 122
□ aged ―― 162
□ age-dependent ―― 116
□ agent ―― 27, 62, 95, 126, 129, 178, 196
□ age-related ―― 174
□ aging ―― 124, 174, 234
□ a high degree of～ ―― 26, 190
□ AIDS ―― 61, 186
□ aim ―― 24, 40, 75, 107, 133, 163, 182
□ aimed at～ing ―― 47
□ air ―― 87
□ alcoho ―― 175
□ alcoholics ―― 211
□ alert ―― 154
□ all ―― 111
□ allergen ―― 26
□ allograft ―― 68, 126, 167
□ almost ―― 77, 94, 186
□ alone ―― 143
□ also ―― 154
□ alter ―― 103, 132, 199
□ alteration ―― 174
□ alternative ―― 135, 239
□ alternatively ―― 140
□ Alternatively, ―― 150
□ although ―― 153
□ always ―― 160
□ Alzheimer's disease 56, 130, 158
□ amino acid ―― 128
□ amount ―― 87, 189
□ amphibian ―― 154
□ amplify ―― 231
□ amyloid ―― 95
□ amyloid deposition ―― 116
□ analogous ―― 51
□ analogous to～ ―― 51
□ analysis ―― 30, 83, 102, 132
□ analyze ―― 30, 130, 136, 227
□ an array of～ ―― 108
□ androgen ―― 214
□ androgen receptor ―― 139
□ angiogenesis ―― 108
□ animal ―― 25, 29
□ animal model ―― 30, 173
□ animal study ―― 43
□ annual ―― 187, 234
□ anomaly ―― 63, 111
□ anthrax ―― 75
□ antiapoptotic ―― 25
□ antibiotic ―― 150, 213
□ antibiotics ―― 239
□ antibody ― 22, 83, 89, 109, 151
□ anticancer ―― 126, 231
□ antifungal ―― 95
□ antigen ―― 123, 204
□ antitumor ―― 121
□ antiviral ―― 136, 161
□ a number of～ ―― 61, 69, 175, 239, 244
□ aortic ―― 229
□ apoptosis ―― 144, 193, 206, 235
□ apoptotic ―― 113, 121
□ appear ―― 51
□ appear to～ ―― 51, 81, 129, 196
□ application ―― 28, 217
□ approach ―― 57, 86, 135
□ approach for～ ―― 67
□ appropriately ―― 52
□ approve ―― 161
□ approximately ―― 184
□ approximately equal ―― 87
□ arise ―― 121, 145, 203, 216, 218
□ arise from～ ―― 63
□ aromatase ―― 76
□ arterial pressure ―― 91
□ artery ―― 65
□ arthropod ―― 74
□ article ―― 198
□ as ―― 218
□ as a component of～ ―― 207
□ as a consequence of～ ―― 203

□ as a first step — 208
□ as a marker — 204
□ as a means — 204
□ as a mechanism — 206
□ as an alternative to～ — 209
□ as an example — 208
□ as an index of～ — 209
□ as an inhibitor — 207
□ as a probe — 205
□ as a result of～ — 202
□ as a source of～ — 206
□ as assessed by～ — 138
□ as a substrate — 203
□ as a target — 205
□ as a whole — 206
□ as compared with～ — 192
□ as controls — 205
□ as defined by～ — 198
□ as demonstrated by～ — 195
□ as described — 198
□ as detected by～ — 197
□ as determined by～ — 192
□ as early as～ — 212
□ as early as possible — 169
□ as effective as～ — 214
□ as efficiently as～ — 215
□ as evidenced by～ — 193
□ as expected — 193
□ asexually — 63
□ as few as～ — 215
□ as for～ — 216
□ as high as～ — 213
□ as indicated by～ — 194
□ as judged by～ — 195
□ ask — 41
□ ask whether～ — 41
□ as little as～ — 214
□ as long as～ — 213
□ as low as～ — 212
□ as many as～ — 214
□ as measured by～ — 192
□ as much as～ — 212
□ as observed — 196
□ as of～ — 217
□ as opposed to～ — 157, 196
□ as part of～ — 202
□ aspirin — 209
□ as predicted by～ — 199
□ as reflected by～ — 199
□ as revealed by～ — 198
□ assay — 183, 195
□ assembly — 70
□ assess — 137, 226, 227
□ assessment — 232
□ assess the effect of～ — 57
□ assess the role of～ — 22
□ assess whether～ — 42
□ as shown by～ — 194
□ associated with～ 141, 142, 202
□ association — 136
□ association between～— 123, 139
□ assume — 59
□ assumption — 157
□ a stack of～ — 213
□ as the basis for～ — 210
□ as to～ — 216
□ as well as～ — 211
□ as with～ — 217
□ asymptomatic — 204
□ asynchronous — 197
□ at a concentration of～ — 227
□ at a dose of～ — 231
□ at a rate of～ — 242
□ at high risk for～ — 204, 238
□ at least — 36, 145
□ at least in part — 239
□ attack — 164
□ attempt — 44
□ attendant — 169
□ attenuate — 135
□ at the beginning of～ — 225
□ at the end of～ — 232
□ attributable — 114
□ attributable to～ — 114
□ atypical — 145
□ augment — 100, 185
□ author — 130
□ authority — 36
□ autoimmune — 146
□ autoimmune disease 23, 134, 214

□ autonomic — 199
□ autophosphorylation — 237
□ available — 159, 169, 172, 206, 224
□ available for ～ — 109
□ avian — 197
□ aware — 181
□ axial — 213
□ azide — 185

B

□ bacterial — 230
□ bactericidal activity — 112
□ bacterium — 63, 89, 151, 198
□ barium — 196
□ barrier — 42, 164
□ basal forebrain — 93
□ basal ganglia — 194
□ based on ～ — 145, 177
□ basement membrane — 108
□ basis — 127
□ basis for ～ — 127
□ because — 168
□ because of ～ — 172
□ become — 172, 218
□ begin — 218
□ behavioral — 116, 154
□ believe — 56
□ believed to ～ — 56
□ beneficial — 235
□ besides — 196
□ better than ～ — 236
□ billion — 234
□ bind — 95, 96
□ binding — 81, 100, 141, 195
□ binding site — 50
□ bind to ～ — 68
□ biochemical — 120, 127
□ biochemically — 80
□ biofilm — 35
□ biology — 199
□ biomarker — 135
□ biopsy — 52
□ biotechnology — 172
□ bleeding — 49
□ block — 92
□ blood — 95
□ blood flow — 163
□ blood glucose level — 67
□ blood group antigen — 166
□ blood pressure — 167, 225
□ blood vessel — 31, 241
□ body mass index — 136, 138, 223
□ bone — 129
□ bone formation — 23
□ bone marrow cell — 138
□ bone morphogenetic protein 227
□ border — 78
□ bound to ～ — 128
□ bovine — 22
□ bowel — 196
□ brain — 43, 114, 171
□ branch — 203
□ break — 62
□ breast — 198
□ breast cancer — 28, 44, 65, 76, 206
□ broad — 184
□ bronchial — 142
□ bronchiectasis — 242
□ bronchitis — 242
□ browser — 217
□ budding — 198, 207
□ buffer — 141
□ By contrast, — 151
□ by means of ～ — 238
□ by use of ～ — 245

C

□ calcium — 34
□ calcium influx — 92
□ call — 62
□ called — 25, 62
□ Calmodulin — 69
□ cancer — 35
□ candidate — 24, 130
□ capable of ～ing — 50, 70

- □ capacity —— 36
- □ carbohydrate —— 136
- □ carcinogen —— 85
- □ carcinogenesis —— 85
- □ carcinoma —— 197
- □ cardiac —— 172, 181
- □ cardiovascular disease —— 87, 109
- □ careful —— 225, 240
- □ carefully —— 167
- □ caregiving —— 202
- □ carry out —— 43, 114
- □ cartilage —— 67
- □ catalytic —— 171
- □ catecholamine —— 103
- □ causal —— 196
- □ cause —— 30, 70, 80, 142, 147, 168, 180, 182, 197, 245
- □ cause an increase in～ —— 22
- □ caused by～ —— 211, 217
- □ celiac disease —— 124
- □ cell —— 22
- □ cell cycle —— 129, 230
- □ cell cycle progression —— 56
- □ cell death —— 235
- □ cell growth —— 126, 225
- □ cell polarity —— 132
- □ cell proliferation —— 170
- □ cellular —— 42, 82, 127, 240
- □ central nervous system 64, 103
- □ cerebrovascular —— 209
- □ certain —— 166
- □ cervical —— 239
- □ cervical cancer —— 239
- □ challenge —— 47
- □ change —— 171
- □ change in～ —— 82, 124, 151, 244
- □ channel —— 74
- □ characterization —— 208
- □ characterize —— 133, 231, 235
- □ chemokine —— 34
- □ chemotherapy —— 46, 52, 129, 213, 216, 231
- □ chest —— 187
- □ chlamydial —— 238
- □ Chlamydia pneumoniae —— 27
- □ choice —— 151
- □ cholinergic —— 93
- □ chromium —— 106
- □ chromosome —— 62, 73, 130
- □ chronic —— 123, 242
- □ chronically —— 82
- □ chronically infected —— 82
- □ chronic bronchitis —— 142
- □ chronic rejection —— 58
- □ circadian —— 38, 47
- □ circadian clock —— 125
- □ circulate —— 144
- □ circulation —— 144
- □ circumstance —— 166
- □ circumvent —— 224
- □ class —— 123
- □ clear —— 82, 237
- □ clearly —— 175, 177
- □ clearly defined —— 83
- □ cleft —— 101
- □ clinical —— 112, 124, 143, 150, 208, 239
- □ clinically —— 97, 202
- □ clinical trial —— 130
- □ clock —— 237
- □ closely —— 74
- □ closely related —— 74
- □ cochlea —— 197
- □ cognitive —— 149
- □ cognitive development —— 80
- □ cognitively —— 219
- □ coimmunoprecipitation —— 192
- □ coincident —— 115
- □ coincident with～ —— 115, 171
- □ collectively —— 127
- □ colocalize —— 156
- □ colon —— 37, 85
- □ colon cancer —— 28
- □ colonic —— 49
- □ colorectal cancer —— 53, 86
- □ combat —— 162
- □ combination —— 165
- □ combination therapy —— 165
- □ combined treatment —— 64
- □ commercially available —— 107

□ commitment ―― 75
□ common ―― 114
□ commonly ―― 63
□ common to～ ―― 231
□ communication ―― 153
□ comparable ―― 113
□ comparable to～ ―― 113
□ comparative ―― 83
□ compare ―― 52
□ compared to～ ―― 66
□ compared with～ ―― 52, 193
□ Compared with～ ―― 187
□ compatible ―― 115
□ compatible with～ ―― 115, 147
□ compelling ―― 127
□ compensation ―― 241
□ competence ―― 82
□ complete ―― 202
□ completely ―― 77
□ completely blocked ―― 77
□ complex ―― 70, 81, 83, 122, 217
□ complexed with～ ―― 81
□ complex with～ ―― 144
□ complication ―― 36, 147, 181
□ component ―― 207
□ compound ―― 227
□ comprehensive ―― 130, 203
□ comprise ―― 215
□ computed tomography ―― 113
□ computer-based ―― 150
□ concentration ―― 158, 185, 212, 223, 232
□ concept ―― 195
□ concern ―― 146
□ concerning ―― 175
□ conclude ―― 36
□ conclude that～ ―― 36
□ concomitant ―― 116
□ concomitant with～ ―― 93, 116
□ condition ―― 88
□ conduct ―― 31
□ confirm ―― 111
□ confirm that～ ―― 37
□ conformation ―― 171
□ congenital ―― 62, 111, 226
□ Consequently, ―― 103
□ conserve ―― 73
□ consider ―― 193, 208
□ considerably ―― 230
□ considered to be～ ―― 57
□ consistent with～ ―― 114
□ consist of～ ―― 26, 66
□ constitute ―― 234
□ constitutively ―― 85
□ constitutively active ―― 85
□ contain ―― 81, 137, 226
□ contaminant ―― 208
□ contemporary ―― 208
□ contrary to～ ―― 157
□ contribute ―― 42
□ contribute to～ ―― 42, 55, 125, 172, 174, 233
□ contribution ―― 126
□ control ―― 73, 84, 129, 182, 188, 193, 199
□ controversial ―― 243
□ Conversely, ―― 102
□ conversion ―― 23
□ cooperativity ―― 228
□ coordinately ―― 31
□ copper ―― 106
□ coronary heart disease ―― 182, 204
□ correlate ―― 53, 216
□ correlated with～ ―― 53
□ correlation ―― 124
□ correlation between～ ―― 124
□ corresponding ―― 88
□ cortically ―― 82
□ corticosterone ―― 170
□ cost ―― 234
□ country ―― 75, 202
□ critical ―― 57, 140, 175
□ critical for～ ―― 122
□ crop ―― 25
□ cross ―― 106
□ cross-sectional ―― 196
□ crucial ―― 108
□ crucial for～ ―― 108
□ crystal structure ―― 227
□ cure ―― 169

□ current ——— 24, 208
□ currently ——— 61
□ currently available ——— 151
□ cutaneous ——— 196
□ cycle ——— 171
□ cyclin ——— 56
□ cystic fibrosis ——— 66
□ cytokine ——— 100, 138, 151
□ cytologic ——— 239
□ cytology ——— 187
□ cytoplasm ——— 29
□ cytoplasmic ——— 156
□ cytotoxic T-lymphocyte ——— 141
□ cytotoxin ——— 121

D

□ damage ——— 64, 144, 194
□ damaged ——— 175
□ data ——— 31
□ data for～ ——— 204
□ data from～ ——— 136
□ data on～ ——— 139
□ deal with～ ——— 218
□ death ——— 40, 133, 146, 182, 238
□ decade ——— 54, 68
□ decision ——— 239
□ decline ——— 239
□ decrease ——— 34, 170
□ decreased ——— 162, 194
□ decrease in～ ——— 34, 128
□ defect ——— 82, 88
□ defective ——— 42
□ defense ——— 42
□ defense against～ ——— 121, 125
□ deficiency ——— 110, 226
□ define ——— 194
□ degradation ——— 25, 108
□ degree ——— 164
□ delay ——— 245
□ dementia ——— 194, 202
□ demographics ——— 84
□ demonstrate ——— 22, 192
□ demonstrate that～ ——— 35, 102, 174, 240
□ demonstrate the presence of～ 23
□ depend ——— 36
□ dependence ——— 141
□ dependence on～ ——— 141
□ dependent ——— 76
□ dependent on～ ——— 111
□ depend on～ ——— 36, 66
□ depression ——— 179, 202, 214
□ deprivation ——— 165
□ derive ——— 49
□ derived from～ ——— 49
□ descendant ——— 193
□ describe ——— 231
□ design ——— 58, 130
□ design to～ ——— 58
□ despite～ ——— 158
□ destruction ——— 89, 243
□ detail ——— 196
□ detect ——— 22, 103, 114
□ detected in～ ——— 87
□ determination ——— 132, 218
□ determine ——— 23, 127, 205
□ determine the effect of～ ——— 23
□ determine whether～ ——— 65, 107
□ detoxify ——— 29
□ deuterostome ——— 74
□ develop ——— 69, 127, 135, 169, 172, 202, 232
□ developed ——— 29
□ developing country ——— 73
□ development ——— 41, 57, 64, 79, 99, 196, 233, 245
□ developmental ——— 109
□ developmentally ——— 78
□ developmentally regulated ——— 78
□ dextrose ——— 128
□ diabetes ——— 142
□ diagnose ——— 52
□ diagnosis ——— 35, 122, 173
□ diagnostic ——— 109
□ diagnostic test ——— 109, 112, 169
□ diarrhea ——— 238
□ diarrheal ——— 73

□ die ―― 164
□ die of～ ―― 162
□ dietary ―― 136
□ differ ―― 63
□ difference ―― 65, 84
□ difference between～ ―― 124
□ difference in～ ―― 137, 223, 233
□ different ―― 185, 229
□ different from～ ―― 110
□ differential ―― 232
□ differential diagnosis ―― 161, 176
□ differentially ―― 75
□ differentially expressed ―― 75
□ differentiate ―― 65
□ differentiate into～ ―― 65
□ differentiation ―― 131
□ differentiation pathway ―― 244
□ differ from～ ―― 63
□ difficult ―― 218
□ difficulty ―― 194
□ directly ―― 91
□ directly by～ ―― 91
□ disagreement ―― 156
□ discontinuous ―― 125
□ discover ―― 49
□ discovery ―― 124
□ discuss ―― 229
□ disease ―― 61, 70, 73, 100, 168, 174, 187, 209, 210, 225, 234, 238
□ disease pathogenesis ―― 173
□ disease resistance ―― 25
□ display ―― 141, 187
□ distant metastasis ―― 108
□ distinct ―― 41, 233
□ distinct from～ ―― 110
□ distinctly ―― 88
□ distinguish ―― 169
□ distribute ―― 73
□ DNA binding ―― 107, 223
□ DNA sequence ―― 26
□ doctor ―― 169
□ domain ―― 66, 237
□ dominant-negative ―― 79
□ dopamine ―― 136
□ dose ―― 66, 185, 245
□ dose-dependent ―― 212
□ double-blind ―― 114
□ Down's syndrome ―― 42
□ downstream ―― 67
□ dramatically ―― 185
□ drastic ―― 197
□ drive ―― 171
□ drowsiness ―― 48
□ drug ―― 112
□ drug dependence ―― 154
□ drug dose ―― 111, 159
□ drug interaction ―― 189
□ drug response ―― 142, 161, 207
□ drug therapy ―― 163
□ due to～ ―― 135, 173, 239
□ duration ―― 240
□ during the course of～ ―― 230
□ during the process of～ ―― 240
□ dynamic ―― 184
□ dynamics ―― 224
□ dysfunction ―― 39, 125, 172
□ dystrophic ―― 232

E

□ each ―― 162
□ each other ―― 229
□ earlier ―― 198
□ earliest ―― 231
□ early ―― 57, 154, 158, 165
□ echocardiography ―― 138
□ economically ―― 25
□ effect ―― 92, 131, 195, 208, 235
□ effective ―― 25, 80, 131,152, 159, 165
□ effect of … on～ ―― 129, 168
□ effect on～ ―― 43, 100, 144
□ effector ―― 70
□ efficacy ―― 192, 216, 224
□ efficient ―― 42
□ ejection fraction ―― 138
□ elemental ―― 95
□ elevate ―― 193

□ elimination ——— 115
□ ELISA ——— 193
□ elongation ——— 22
□ elucidate ——— 134
□ elucidate the mechanism ——— 23
□ embark ——— 208
□ embryo ——— 243
□ embryogenesis ——— 62
□ embryonic ——— 65, 164
□ emphasis ——— 167
□ emphasize ——— 142
□ enable ——— 30, 205
□ encode ——— 55, 140
□ endocrine ——— 189
□ endogenous ——— 36
□ endometrial ——— 52
□ endothelial ——— 138
□ endothelial cell ——— 27, 94
□ energy ——— 128
□ engineer ——— 96
□ enhance ——— 50, 138
□ enhanced ——— 185, 237
□ enhancement ——— 232
□ enlargement ——— 198
□ enteric ——— 74
□ environment ——— 125, 147
□ environmental ——— 62
□ environmental factor ——— 193, 217
□ enzymatic ——— 102
□ enzyme ——— 29, 80, 109, 120, 129, 141, 171, 208, 228
□ epidemic ——— 123
□ epidemiologic ——— 37
□ epidemiologic study ——— 37
□ epidermal ——— 107
□ epidermal growth factor receptor ——— 51
□ epithelial ——— 28, 75, 197, 235
□ epithelial cell ——— 100, 164
□ epithelium ——— 63
□ equal ——— 87
□ error ——— 62
□ essential ——— 106
□ essential for～ ——— 106
□ establish 58, 154, 177, 207, 229
□ estimate ——— 36, 234
□ estimate that～ ——— 36
□ estrogen ——— 76
□ estrogen receptor ——— 155, 226
□ estrogen replacement ——— 214
□ ethnic ——— 233
□ ethnically ——— 73
□ ethnic group ——— 111
□ ethnicity ——— 138
□ etiologic agent ——— 62
□ etiology ——— 24, 193
□ evaluate ——— 58, 85, 123, 126, 224
□ evaluate the effect of～ ——— 24
□ evaluate whether～ ——— 42
□ evaluation ——— 109, 112
□ evasion ——— 32
□ even if～ ——— 176
□ event ——— 47, 67, 180
□ even though ——— 153
□ evidence ——— 194, 198, 206, 233, 234, 240
□ evidence ——— 49
□ evidence-based medicine ——— 210
□ evidence for～ ——— 127
□ evidence that～ ——— 199, 236
□ evident ——— 35, 109
□ evoke ——— 168
□ evolution ——— 140, 231, 237
□ exacerbate ——— 187
□ exaggerate ——— 241
□ examination ——— 27, 149
□ examine ——— 155, 182
□ examine the role of～ ——— 24
□ examine whether～ ——— 41
□ excellent ——— 132
□ exception ——— 227
□ excitatory amino acid ——— 86
□ exclusive ——— 157
□ exclusively ——— 94
□ exclusively in～ ——— 94
□ excrete ——— 95, 218
□ excretory ——— 205
□ exercise ——— 244
□ exercise tolerance testing ——— 204
□ exert ——— 43, 100

□ exhibit — 66, 121, 184, 185, 197
□ exist — 65
□ exist in ～ — 65
□ expand — 242
□ expect — 54
□ expected for ～ — 110
□ expected to ～ — 54, 79, 203
□ experiment — 224, 232
□ experimental — 84
□ experimentally — 83
□ experimentally determined — 83
□ explain — 28, 92, 157, 177
□ explanation — 134
□ explanation for ～ — 134
□ explore — 168
□ exposure — 43, 48, 196
□ exposure to ～ — 41
□ express — 85, 93, 94, 100, 155
□ expressed in ～ — 74
□ expressed on ～ — 51
□ expression 97, 113, 195, 197, 244
□ expression by ～ — 126
□ extensive — 158
□ extensively — 82
□ extensively studied — 82
□ external — 125, 172
□ extracellular — 223
□ extracellular matrix — 50
□ extremely — 112

F

□ facilitate — 102
□ fact — 177
□ factor — 30, 81, 93, 156, 172
□ fail — 69
□ fail to ～ — 69
□ fall — 211, 225
□ family — 177
□ family history — 225
□ fat — 53
□ fatty acid synthase — 77
□ feature — 111, 143
□ fecal — 73, 137
□ feces — 95
□ feeding — 185
□ female — 89, 186
□ femur — 65
□ fertilize — 69
□ fetal — 35, 62, 75, 103
□ few — 43
□ FGF — 92
□ fibril — 95
□ fibroblast — 50, 164
□ fibrosis — 28
□ film — 187
□ filtrate — 49
□ Finally, — 99
□ find — 53, 100
□ finding — 28, 34, 114, 115, 129, 134, 146, 170, 225, 229, 234, 237
□ fine — 87
□ fission yeast — 132
□ flow cytometry — 198
□ fluorescence — 183
□ fluorine — 106
□ focus — 67
□ focus on ～ — 67
□ follow — 244
□ followed by ～ — 48
□ follow up — 213
□ food — 122
□ food intake — 77, 188
□ for example — 233
□ for instance — 236
□ form — 94, 144, 152, 214, 226, 228, 236
□ formate — 185
□ formation — 32, 35, 153, 231
□ former — 186
□ for the first time — 244
□ for the purpose of ～ — 241
□ found to ～ 51, 85, 87, 95, 96, 97, 107, 111, 137, 138, 213
□ fragment — 96
□ frail — 218
□ frequent — 211
□ frequently — 228

□ full —— 78
□ fully —— 81
□ function — 35, 43, 70, 140, 143, 173, 209, 226, 234
□ functional —— 50, 114, 150, 216
□ functioning —— 155
□ fundamental —— 166
□ fungus —— 63
□ further —— 22, 42, 129, 227
□ Further, —— 164
□ furthermore, —— 164
□ future —— 99, 130

G

□ gain —— 126
□ galanin —— 93
□ gastric —— 193
□ gastroenteritis —— 49
□ gastrointestinal —— 216
□ gene —— 24
□ gene expression —— 28, 88, 93, 184, 205, 223, 240
□ gene for～ —— 130
□ general —— 159
□ generally —— 59, 245
□ generate —— 25, 130
□ generation —— 168
□ gene-specific —— 195
□ gene therapy —— 99
□ genetic —— 126, 132, 177, 199
□ genetically —— 146, 233
□ genetically engineered —— 79
□ genetic disease —— 139
□ genetic factor —— 142, 147, 206
□ genetic interaction —— 130
□ genital —— 91, 164
□ genital herpes —— 29
□ genome —— 217
□ genomic —— 38
□ gestation —— 41
□ Given —— 177
□ give rise to～ —— 235
□ gland —— 189
□ glucosamine —— 235
□ glucose —— 212
□ glutamate receptor —— 92
□ glycogen synthase kinase — 227
□ G protein —— 66
□ G protein-coupled receptor — 23
□ graft —— 22
□ great —— 47, 132
□ greater —— 37
□ greater than～ —— 181
□ group —— 81, 84, 112, 137
□ growth —— 103, 115, 197, 199
□ growth factor —— 50, 92
□ growth hormone secretagogue — 236
□ GTPase —— 66

H

□ haplotype —— 236
□ haptoglobin —— 144
□ harmful —— 166
□ have a role in～ —— 25
□ have the potential to～ —— 25
□ HBV —— 204
□ head —— 180
□ headache —— 48, 64
□ health —— 106, 122
□ healthy —— 136, 137, 151
□ heart —— 74, 189
□ heart disease —— 133
□ heart failure —— 127
□ helical —— 66
□ help —— 112
□ helper T cel —— 166
□ help to～ —— 112, 234
□ hemodialysis —— 209
□ hemoglobin —— 144
□ hence —— 161
□ hepatitis —— 175
□ here —— 42
□ hereditary —— 225
□ heritable —— 172
□ hierarchical —— 114
□ high —— 26, 103

□ higher ——— 66
□ higher than～ ——— 181
□ highlight ——— 129
□ highly conserved ——— 73
□ highly susceptible ——— 85
□ hink ——— 56
□ hippocampal ——— 214
□ Hispanic ——— 186
□ histologic ——— 232
□ histologically ——— 189
□ histology ——— 113
□ histone ——— 56
□ HIV ——— 40, 109, 164
□ hold ——— 157
□ homolog ——— 207
□ homology ——— 143
□ homology to～ ——— 143
□ homozygous ——— 109
□ homozygous for～ ——— 109
□ hopefully ——— 131
□ hormone ——— 181, 205, 226
□ hormone receptor ——— 51
□ hospital ——— 234
□ host ——— 125
□ host-derived ——— 122
□ However, ——— 148
□ human ——— 24, 68, 100
□ human herpesvirus ——— 91
□ human immunodeficiency virus ——— 53
□ human leukocyte antigen ——— 123
□ humans ——— 75
□ human T-cell leukemia virus ——— 97
□ hyaluronan ——— 137
□ hybridization ——— 232
□ hydrophobic ——— 26
□ hygiene ——— 73
□ hypothalamic ——— 38
□ hypothesis ——— 58, 101
□ hypothesize ——— 38

I

□ ideal ——— 166
□ identical ——— 113
□ identical to～ ——— 113
□ identifiable ——— 123
□ identification ——— 139
□ identified as～ ——— 106
□ identified in～ ——— 130
□ identify ——— 32, 44, 47, 75, 112, 132, 169, 204, 238
□ If ——— 176
□ illness ——— 161
□ image ——— 213
□ imaging ——— 173, 196
□ immediate ——— 34, 36
□ immediately ——— 244
□ immobilization ——— 244
□ immune ——— 82
□ immune cell ——— 241
□ immune response ——— 166, 185
□ immune system ——— 162
□ immunization ——— 202
□ immunochemical ——— 41
□ immunocompromised patient ——— 70
□ immunocytochemistry ——— 197
□ immunodeficient ——— 121
□ immunosuppressive ——— 91
□ immunosuppressive therapy ——— 165
□ impact ——— 139
□ impact on～ ——— 139
□ impair ——— 142, 154, 219
□ impairment ——— 194
□ implicate ——— 207
□ implicated in～ ——— 50
□ implication ——— 129
□ implication for～ ——— 130
□ imply ——— 34
□ imply that～ ——— 34
□ important ——— 25, 35, 77, 129, 193
□ important for～ ——— 22, 107
□ improve ——— 64, 79, 99, 116
□ improvement ——— 73, 188
□ in accordance with ——— 170
□ inactive ——— 227
□ In addition, ——— 165
□ in addition to～ ——— 172
□ in agreement with～ ——— 170
□ in a manner similar to～ ——— 238

□ in an attempt to～ ——— 224
□ in an effort to～ ——— 231
□ in association with～ ——— 145
□ incidence ——— 89, 106, 135, 194
□ in close proximity to～ ——— 241
□ include ——— 37
□ in combination with～ — 226, 245
□ in common with～ ——— 226
□ in comparison with～ ——— 188
□ in complex with～ ——— 227
□ In conclusion, ——— 175
□ inconclusive ——— 139
□ in conjunction with～ ——— 228
□ in contact with～ ——— 229
□ In contrast, ——— 151
□ in contrast to～ ——— 156
□ incorporation ——— 138
□ incorporation into～ ——— 138
□ increase — 33, 54, 97, 165, 189
□ increased ——— 106, 116, 142, 170, 173
□ increased ～-fold ——— 185
□ increase in～ ——— 33, 116, 188
□ increase the rate of～ ——— 26
□ increasing ——— 194, 240
□ indeed ——— 165
□ independent ——— 180
□ independent of～ ——— 111, 147
□ indicate ——— 26
□ indicate that～ — 28, 34, 92, 233
□ indicate the presence of～ — 26
□ indication ——— 162
□ indicator ——— 238
□ indispensable ——— 225
□ indistinguishable ——— 216
□ indistinguishable from～ 97, 110
□ individual ——— 80, 157, 230
□ induce ——— 146, 153, 163, 244
□ induced by～ ——— 113
□ induce the expression of～ — 27
□ industrial ——— 48
□ ineffective ——— 166
□ inefficient ——— 31
□ in excess of～ ——— 234
□ in fact ——— 97, 166
□ in favor of～ ——— 234, 244
□ infect ——— 40
□ infected ——— 27, 100, 162, 163
□ infected with～ ——— 40, 53
□ infection — 30, 97, 125, 142, 149, 164, 204, 238, 241
□ infectious ——— 49
□ infer ——— 50
□ inferred from～ ——— 50
□ inflammation ——— 228
□ influence ——— 182, 200
□ influence on～ ——— 141
□ influenza ——— 238
□ information ——— 120
□ information about～ ——— 120
□ information on～ ——— 140
□ infusion ——— 128
□ in good agreement with～ — 224
□ inherent ——— 135
□ inhibit ——— 151, 166
□ inhibition ——— 83, 121, 212
□ inhibition by～ ——— 126
□ inhibitor — 78, 79, 101, 209, 227
□ in humans ——— 106
□ initiation ——— 156
□ injection ——— 181
□ injure ——— 68
□ injury ——— 180, 211, 240, 241
□ in light of～ ——— 237
□ innate immune system ——— 125
□ inner ——— 88
□ inner ear ——— 88
□ in order to～ ——— 126
□ in part ——— 235
□ in place of～ ——— 239
□ in proportion to～ ——— 240
□ in response to～ ——— 132, 164
□ insight ——— 125, 126
□ in spite of～ ——— 158
□ Instead, ——— 150
□ insulate ——— 125
□ insulin ——— 36, 68, 102, 185, 227
□ insult ——— 121
□ In summary, ——— 174
□ intake ——— 53, 136, 146

□ integral membrane protein — 55
□ integration — 38
□ integron — 198
□ intense — 161
□ interact — 69
□ interaction — 127
□ interaction between ～ — 85, 175
□ interact with ～ — 69, 192
□ Interestingly, — 100
□ interface — 35
□ interfere — 69
□ interfere with ～ — 70
□ interferon — 52, 100
□ interindividual — 206
□ interleukin — 121
□ intermediate — 24
□ in terms of ～ — 157
□ interplay — 150, 217
□ intestinal — 228
□ in the absence of ～ — 223
□ in the case of ～ — 225
□ in the context of ～ — 31, 229
□ in the face of ～ — 234
□ in the form of ～ — 235
□ in the future — 172
□ in the presence of ～ — 240
□ intracellular — 34, 69, 122, 150
□ intraneuronal — 133
□ intraperitoneal — 181
□ intrinsic — 101, 203
□ intrinsically — 123
□ introduce — 150, 206
□ investigate — 22, 126, 233
□ investigate the effect of ～ — 27
□ investigate whether ～ — 40
□ investigation — 68, 125
□ in view of ～ — 245
□ in vitro — 27, 115, 158
□ in vivo — 23, 43, 68, 115, 158
□ involve — 47
□ involved in ～ — 47, 56, 57, 69, 199, 225, 236
□ iodine — 106, 146
□ ion — 81
□ ionic strength — 77, 141
□ iritis — 228
□ iron — 106, 181, 189
□ ischemia — 164, 168
□ ischemic — 138
□ ischemic injury — 235
□ isoflavone — 129
□ isoform — 140
□ it … accepted that ～ — 59
□ it … assumed that ～ — 59
□ it is suggested that ～ — 103

K

□ key — 56
□ kidney — 37, 93, 205, 235
□ kinase — 23, 237
□ knockdown — 170
□ know — 86
□ known — 30
□ known as ～ — 69
□ known to ～ — 25, 57, 61, 77, 143, 153, 189, 204, 228

L

□ laboratory test — 159
□ lack — 63, 110, 173
□ largely — 68, 143
□ larger — 87
□ last — 41
□ lasting — 43
□ latent — 48
□ late stage — 154
□ latter — 186
□ latter half — 171
□ lead — 37, 48, 105
□ leading — 180
□ lead to ～ — 48, 67, 79, 128, 131, 225, 237
□ lean — 77
□ learn — 46
□ learned about ～ — 46
□ leprosy — 125
□ leptin — 185

□ lesion 51, 134
□ less than ～ 180
□ lethal 30
□ leukemia 236
□ leukemia inhibitory factor 101
□ level 76
□ liberate 144
□ ligand 85
□ likewise 167
□ limit 41, 73, 163
□ limited 162, 177, 216
□ lineage 75
□ lineage commitment 76
□ linearly 97
□ linearly with ～ 97
□ lines of evidence 233
□ link 123, 124
□ link between ～ 124
□ linked to ～ 123
□ lipid bilayer 48
□ littermate 111, 212
□ little 144
□ little is known about ～ 153
□ live 75
□ liver 27
□ liver disease 159
□ liver failure 58
□ living 115
□ local 108, 241
□ localize 131
□ localized 217
□ locate 51
□ located in ～ 51
□ location 205
□ locus 156
□ long 203
□ longitudinally 213
□ long-term 158
□ loop 77
□ lose 38
□ loss 116, 141, 226
□ loud 24
□ low 185
□ lower 91
□ lower than ～ 181
□ lung 41
□ lung cancer 131, 187, 240
□ lupus erythematosus 180
□ Lyme disease 123
□ lymphangioleiomyomatosis 94
□ lymphoma 51

M

□ machinery 207
□ macrophage 31
□ magnesium 81
□ magnetic resonance imaging 194
□ main 46
□ mainly 95
□ mainly in ～ 95
□ maintain 77, 223
□ major 111, 149, 205
□ major histocompatibility complex 140
□ make it possible to ～ 150, 213
□ male 89, 186
□ malignancy 245
□ malignant 28, 63, 108, 235
□ mammalian 46, 77, 154
□ mammary 235
□ manganese 106
□ manifestation 238
□ manner 88, 114
□ map 130
□ MAP kinase 38
□ mapping 50
□ markedly 187, 193
□ markedly reduced 76
□ marker 123
□ maternal 23, 35
□ matter 87
□ maturation 88, 114
□ maximal 171
□ maximum 212
□ mean 122
□ mean age 198
□ measles 202
□ measure 167

□ mechanism — 38
□ mechanism by which ～ — 125
□ mediate — 56
□ mediated — 82
□ medical — 233
□ medical care — 233
□ medication — 189
□ melanoma — 28, 145, 196
□ member — 38
□ membrane — 95, 168
□ membrane protein — 77
□ memory T cell — 68
□ menopausal — 180
□ menopause — 182
□ mental — 149
□ mercury — 95
□ metabolic — 101, 147, 157
□ metabolism — 31, 103, 172
□ metallothionein — 128
□ metastatic — 134
□ method — 113, 131, 132, 206, 210, 213
□ method for ～ — 127, 232
□ MHC — 126
□ microbial — 208
□ microorganism — 32, 241
□ microtubule-associated protein — 139
□ microvascular endothelial cell 68
□ migraine headache — 37
□ migration — 78
□ million — 36
□ minicircle — 96
□ mismatched — 166
□ mite — 61
□ mitochondria — 174
□ mitogen-activated protein kinase — 115
□ mitral — 229
□ mobilization — 138
□ model — 108
□ model for ～ — 23
□ modification — 227, 240
□ modulate — 34, 55, 175
□ molecular — 48
□ molecular basis — 140
□ molecular dynamics — 224
□ molecular function — 237
□ molecular interaction — 48
□ molecular mechanism — 131
□ molecule — 27, 173
□ monitor — 67
□ monitoring — 161
□ monoclonal — 59
□ monoclonal antibody — 107, 143
□ more — 131
□ more likely to ～ — 82
□ Moreover, — 165
□ more than ～ — 179
□ mortality — 87, 165, 188
□ mortality rate — 66
□ mosaic — 32
□ most — 70, 161
□ most likely — 37
□ motif — 143
□ motor — 233
□ mouse — 66
□ mouse model — 24
□ mRNA — 25
□ much — 46
□ much greater — 228
□ much larger — 87
□ multiple — 171
□ multiple sclerosis — 217
□ murine — 164
□ muscle — 102, 172
□ muscle mass — 244
□ mutant — 30, 79, 97, 116, 132, 134, 197, 206
□ mutate — 88
□ mutation — 24, 26, 34, 39, 73, 102, 163, 237
□ mutually — 157
□ mutually exclusive — 88
□ mycoplasmal — 98
□ Mycoplasma pneumoniae — 97
□ myocardium — 138
□ myogenesis — 81

N

□ nausea — 48
□ nearly — 113
□ necessary — 107
□ necessary for ～ — 78, 107
□ necrosis — 235
□ need — 129, 161, 178, 227
□ need for ～ — 129
□ need to ～ — 167
□ negative — 223
□ negatively — 93
□ negative-strand RNA — 135
□ nematode — 74
□ nerve growth factor — 192
□ nerve terminal — 136
□ nervous system — 166, 174
□ network — 34
□ neural — 199
□ neural crest — 57
□ neural tube — 207
□ neuroendocrine — 199
□ neurofibroma — 155
□ neuroma — 24
□ neuron — 153, 197, 233
□ neuronal — 33
□ neuronal cell — 207
□ neuropeptide — 93
□ neuroscience — 166
□ neurotropic — 91
□ Nevertheless, — 149
□ new — 27
□ newborn infant — 111
□ newly — 135
□ newly synthesized — 76
□ next — 54
□ nicotine — 43
□ noise — 24
□ Nonetheless, — 149
□ noninvasive — 113
□ nonmammalian — 166
□ normal — 48, 137, 149, 197
□ normal cell — 31
□ normally — 229
□ note — 35
□ note that ～ — 35
□ not fully understood — 81
□ not only ～ but also … — 103
□ novel — 29, 57, 86, 130, 140, 172, 183, 217, 232
□ now — 109
□ N-terminal — 110
□ nuclear hormone receptor — 226
□ nuclear localization — 107
□ nucleotide — 81
□ nucleus — 29
□ nude mouse — 235
□ number — 116
□ numerous — 139

O

□ obese — 102
□ obesity — 123, 188, 236
□ objective — 65
□ observation — 26, 115
□ observational — 136
□ observation that ～ — 145
□ observe — 82, 88, 138, 184, 212
□ observed for ～ — 49
□ observer — 156
□ obstruction — 196
□ obstructive sleep apnea — 146
□ obtain — 113, 213, 224
□ obtained from ～ — 164
□ occur — 41, 94, 97, 142, 226, 228, 240
□ occur during ～ — 171
□ occurrence — 35
□ occur through ～ — 153
□ offer — 127
□ offspring — 231
□ of paramount importance — 32
□ often — 51
□ old — 218
□ older — 179, 236
□ oligonucleotide — 28
□ once — 177
□ only — 155

□ onset — 48, 122, 198, 205
□ on the basis of ～ — 224
□ On the contrary, — 151
□ on the order of ～ — 239
□ On the other hand, — 152
□ oocyte — 69
□ opportunistic infection — 53
□ optimal — 150
□ oral — 74, 137, 224
□ organ — 189, 205
□ origin — 37, 59, 134
□ originate — 64
□ originate from ～ — 64
□ orphan nuclear receptor — 29
□ orphan receptor — 94
□ osteoblast — 192
□ osteoprogenitor — 88
□ other — 112
□ other than ～ — 182
□ outbreak — 49
□ ovarian — 121
□ ovarian cancer — 139
□ ovary — 78
□ overall — 188
□ overreplication — 197
□ overstimulation — 92
□ over the course of ～ — 231
□ owing to ～ — 173
□ oxidation — 128
□ oxidative — 42, 168
□ oxygen — 164
□ oxygenation — 34

P

□ pacemaker — 38, 47
□ pain — 64, 241
□ pancreas — 189
□ pancreatic — 228
□ parenteral — 208
□ Parkinson's disease — 158
□ partially — 80
□ partially purified — 80
□ participate — 64
□ participate in ～ — 64
□ particle — 49
□ particular — 112, 177, 197, 223
□ particularly — 218
□ particulate — 87
□ partnership — 169
□ pass — 35
□ patch — 26
□ pathogen — 74, 162
□ pathogenesis — 55, 236
□ pathogenic — 198
□ pathologic — 37, 111
□ pathology — 187, 203, 239
□ pathway — 25, 31, 64, 132, 199
□ pathway for ～ — 133
□ patient — 49
□ patient with ～ — 36, 38, 82, 109, 130, 209
□ pattern — 28, 88, 113
□ peak — 120, 171
□ penicillin — 239
□ people — 214
□ peptide — 215
□ percent — 155
□ perception — 82
□ perform — 83, 149, 244
□ performed on ～ — 52
□ perfuse — 232
□ perhaps — 93
□ perhaps by ～ — 93
□ period — 48, 62
□ peripheral — 64, 86
□ peripheral vascular resistance — 225
□ per se — 147
□ person — 53
□ per year — 242
□ phagocytosis — 31
□ pharmacopoeia — 208
□ phenotype — 97, 116
□ phorbol estèr — 59
□ phosphate — 81
□ phospholipase — 31
□ phospholipid — 168
□ phosphorylation — 225
□ phosphotyrosine — 102

□ photoreceptor — 232
□ phylogenetic — 163
□ physical — 167, 180
□ physically — 192
□ physician — 35, 150, 154, 160, 169, 176, 218, 237
□ physiologic — 77
□ physiological — 229
□ physiologically — 85, 216
□ physiologically relevant — 85
□ pig — 22
□ pilot — 80
□ pineal gland — 125
□ pituitary — 226
□ place — 167
□ placebo — 192
□ placenta — 106
□ plasma — 128, 170
□ plasma cell — 199
□ platelet — 209
□ platform — 166
□ plausible — 134
□ play an important role in～ — 103, 223
□ play a role in～ — 129
□ pluripotent — 101
□ pneumonia — 162, 238, 242
□ point mutation — 62
□ poisoning — 37
□ poliovirus — 137
□ pollution — 87
□ polymerase — 97
□ polymorphism — 139, 140
□ polyp — 245
□ poorly — 148
□ poorly understood — 67, 148, 173
□ population — 70, 73, 110, 157, 206, 213, 242
□ positive — 109
□ positive for～ — 109
□ positively charged — 77
□ positivity — 89
□ possibility — 22, 160, 176
□ possible — 124
□ possibly — 91
□ possibly by～ — 91
□ postgenomic — 135
□ postnatal — 76
□ posttranslational modification — 108
□ posttreatment — 123
□ postulate — 55
□ postulated to～ — 55
□ potency — 185
□ potent — 209
□ potential — 50, 154, 245
□ potential for～ — 132
□ potentially — 174
□ potentially important — 86
□ potently — 151
□ poxvirus — 230
□ practical — 206
□ practicing physician — 181
□ predict — 55
□ predicted to～ — 55
□ prediction — 47
□ predictor — 180
□ predispose — 146
□ predominantly — 94
□ predominantly in～ — 94
□ preferentially — 96
□ preferentially to～ — 96
□ pregnancy — 225
□ premenopausal — 94
□ prenatal — 41
□ prepare — 52, 70
□ preschool — 49
□ present — 35, 58, 120, 130, 154
□ present evidence — 230
□ preserve — 234
□ presumably — 92
□ presumably by～ — 92
□ presume — 55
□ presumed to～ — 55
□ prevalence — 73
□ preventive — 158
□ previous — 96
□ previously — 29, 122, 177, 195
□ Previously, — 102
□ previously reported — 74
□ primarily — 93

☐ primarily in～ — 93
☐ primarily responsible for～ — 89
☐ primary — 37
☐ primate — 114
☐ priming — 68
☐ prion — 23, 133
☐ proapoptotic — 113, 121
☐ probably — 150, 162, 240
☐ probe — 232
☐ problem — 149, 166
☐ proceed — 114
☐ process — 131, 153, 231, 232
☐ processivity — 77
☐ produce — 62, 192
☐ product — 51, 208
☐ production — 50, 100, 243
☐ profound — 139, 244
☐ progenitor cell — 138, 175
☐ progesterone receptor — 155
☐ progress — 99
☐ progression — 187
☐ prokaryote — 231
☐ proliferation 31, 75, 94, 151, 207
☐ prolonged — 202
☐ prominent — 204
☐ promising — 67
☐ promote — 32
☐ promoter — 230
☐ promoter activity — 78
☐ prompt — 135
☐ property — 101, 141, 157
☐ prophylaxis — 75
☐ proportion — 26
☐ propose — 29, 159
☐ prostate cancer — 92, 130
☐ protease — 89, 133
☐ protect — 61
☐ protect against～ — 61, 238
☐ protection — 121
☐ protection against～ — 121
☐ protein — 50
☐ protein kinase — 59, 85
☐ protein-protein interaction — 55
☐ protein structure — 47
☐ proteinuria — 111
☐ prove — 80
☐ provide 120, 125, 127, 134, 140, 164, 166, 183, 236
☐ provide a mechanism — 28
☐ provide evidence — 28
☐ provoke — 208
☐ proximal — 49
☐ psychological — 241
☐ puberty — 93, 198
☐ publish — 233
☐ purify — 80
☐ purulent — 242
☐ putative — 100
☐ put forward — 101

Q

☐ quantitative — 127
☐ quantitatively — 57
☐ questionnaire — 238

R

☐ race — 133
☐ racial — 233
☐ radiation — 196
☐ radiography — 196
☐ radiotherapy — 216
☐ raise — 163, 190
☐ raise the possibility that～ — 77
☐ raise the question of～ — 242
☐ randomized — 114
☐ range — 184
☐ rapid — 113
☐ rapid eye movement — 157
☐ rapidly — 96
☐ rapidly induced — 82
☐ rare — 225
☐ rat — 75
☐ rate — 33, 181, 186, 239
☐ rate-limiting — 176
☐ rather than～ — 82, 180
☐ rationally — 177
☐ rat model — 58

□ react — 70
□ reaction — 41, 176
□ reactive oxygen species 122, 168
□ react with～ — 70
□ readily — 106
□ rearrangement — 243
□ reason — 242
□ receive — 137
□ recent — 131, 237
□ recently — 233
□ Recently, — 101
□ receptor — 56, 68, 86, 110, 227, 236
□ receptor for～ — 59
□ recipient — 167
□ recognize — 158
□ recombinant — 79, 96
□ recombination — 26
□ recommend — 245
□ recommendation — 159
□ record — 67
□ red cell — 128, 144
□ reduce — 28
□ reduced — 141
□ reduce the number of～ — 29
□ reduction — 116
□ reduction in～ — 138, 197
□ regard — 46, 238
□ regarded as～ — 46
□ regarding～ — 158
□ regeneration — 67
□ region — 101, 130, 199
□ regulate 31, 74, 78, 86, 93, 103
□ regulate the expression of～ — 29
□ regulation — 223
□ regulator — 56
□ regulatory — 81, 181
□ rejection — 126, 149
□ relate — 162
□ related to～ — 162
□ relationship — 122
□ relationship between～ — 122
□ relative — 205
□ relatively — 103, 144
□ relatively little — 86
□ relative to～ — 188
□ relaxation — 163
□ release — 136
□ release from～ — 136
□ relevant — 39
□ relevant to～ — 39
□ relieve — 163
□ rely on～ — 67
□ remain — 68, 74, 82, 101, 136, 143, 148, 154, 173, 175
□ remarkable — 27, 121
□ remarkably — 88
□ remarkably similar — 88
□ removal — 197
□ remove — 245
□ REM sleep — 157
□ renal — 111
□ renal artery — 203
□ renal cell carcinoma — 177
□ renal disease — 159
□ renal dysfunction — 240
□ renal failure — 209
□ renewal — 107
□ repair — 67, 240
□ repeated — 138
□ repertoire — 141
□ replacement therapy — 80
□ replicate — 215
□ replication — 122, 135
□ report — 42, 96, 109, 122, 177, 196
□ reported to～ — 53, 152, 165, 179, 186, 207
□ report the identification of～ — 29
□ represent — 198
□ reproduce — 63, 214
□ reproducible — 113
□ require — 210, 240, 243
□ required for～ — 48, 181
□ required to～ — 42
□ requirement — 128
□ requirement for～ — 128
□ research — 158, 243
□ reservoir — 61
□ residue — 26, 73, 100, 107, 215
□ resistance — 140, 163, 228, 245

□ resistance to～ 142, 213
□ resistant 133
□ resistant to～ 112
□ resorption 83
□ respiratory 63, 161
□ respond 96
□ respond to～ 68, 125, 236
□ response 55, 136, 181, 185, 206, 212
□ response to～ 230
□ responsible 105
□ responsible for～ 83, 105, 152
□ responsive 52
□ resting 163
□ restoration 244
□ result 22
□ result from～ 62
□ result in～ 64, 146, 170, 181, 188
□ retain 149
□ retain the ability to～ 30
□ retinal 184
□ retinal damage 67
□ retinal pigment epithelium 234
□ retinoic acid 103
□ reveal 30
□ reveal that～ 37
□ reveal the presence of～ 30
□ reverse 214
□ review 66, 206, 233
□ revolution 172
□ rhesus monkey 143
□ rheumatoid arthritis 186, 192
□ rise 135
□ risk 37, 133, 162, 189, 240
□ risk factor 44, 87
□ risk for～ 53, 169
□ RNA-binding 143
□ RNA interference 25
□ RNA silencing 70
□ rodent 188
□ rodent model 188
□ role 81, 142, 199, 206, 229
□ rough endoplasmic reticulum 77
□ rule out 176

S

□ Sabin vaccine 137
□ safe 208
□ saliva 91
□ same 154
□ sample 137
□ sample from～ 137
□ SARS 61
□ scavenger 186
□ scientific 210
□ scientist 47
□ screen 163
□ screening 187, 239
□ search 135
□ search for～ 135
□ secondary 207
□ secretion 50, 76, 91, 110
□ secretory 36
□ sedative 211
□ see 211
□ seek 23
□ seem to～ 78, 110, 207
□ segmental 203
□ select 141, 163, 205
□ selective 227
□ selectively 96
□ selectively to～ 96
□ selenium 106
□ senile dementia 38
□ sensitive 64
□ sensitive to～ 112
□ sensitivity 124, 173, 183
□ sensitivity to～ 126
□ sequence 195, 232
□ sequence homology 26
□ sequencing 30
□ serine 74
□ serine/threonine kinase 74
□ serious 61, 180
□ serologic 124
□ seroprevalence 97
□ serotonin 136
□ serum 112, 137
□ serum cholesterol 223

□ serum sample — 22
□ serve — 34
□ serve as～ — 61, 86, 132, 203, 204, 205
□ set up — 217
□ several — 30
□ severe — 109, 179
□ severely — 80
□ severely impaired — 80
□ severity — 134
□ sex — 133
□ sex-matched — 110
□ sexually — 63
□ shed light on～ — 237
□ shortly — 241
□ shortly after～ — 241
□ short-term — 80
□ show — 116, 143, 184, 188, 193, 194, 208
□ shown to～ — 175, 192
□ show that～ — 33, 102, 244
□ shuttle — 29
□ sign — 29, 154
□ signal — 110
□ signaling — 57, 170
□ signaling pathway — 34, 38, 47, 150
□ signaling through～ — 141
□ signal peptide — 110
□ signal transduction — 48, 69
□ signal transduction pathway — 180, 225
□ significant — 42, 99, 100, 133, 138, 151, 180, 188, 202, 223, 245
□ significantly — 66, 110, 230
□ significantly reduced — 73
□ similar — 73, 184
□ similarity — 143
□ similarity to～ — 143
□ similarly — 166
□ similar to～ — 39
□ simulation — 224
□ simultaneously — 97
□ simultaneously with～ — 98
□ since — 168
□ single — 181
□ single-stranded — 203
□ site — 37, 229
□ size — 242, 245
□ skeletal muscle — 101
□ sleep — 146, 171
□ sleepiness — 171
□ slightly — 89
□ slightly higher — 89
□ small — 215
□ smallpox — 181
□ smallpox vaccination — 181
□ smoking — 87, 240
□ smooth muscle — 94, 163
□ social — 154
□ society — 122
□ sodium — 74
□ sole — 59
□ solely — 92
□ solely by～ — 92
□ sometimes — 202
□ so that～ — 169
□ sought to～ — 23, 70, 85, 126
□ source — 128
□ spasm — 163
□ special — 216
□ specialize — 52
□ specific — 26, 28, 96, 156, 182, 213
□ specifically — 95
□ specifically to～ — 95
□ specific for～ — 107
□ specificity — 102, 120, 124
□ specificity for～ — 135
□ specify — 233
□ specimen — 35, 109, 218
□ speculate — 39
□ spheroid — 116
□ spike — 33
□ spleen — 212
□ splice — 140
□ spontaneous — 106
□ spontaneous abortion — 106
□ spontaneously — 143
□ sporadic — 65, 94, 145
□ spread — 74

□ sputum 187, 242
□ stable 223
□ stably 76
□ stably transfected 76
□ stage 69, 231, 244
□ state 157
□ statistically 84
□ statistically significant 84
□ status 124, 149, 180
□ stem cell 75, 107, 124, 243
□ step 57
□ step in ～ 57
□ stimulation 50, 120, 223
□ stimulus 132
□ stool 49
□ strain 112
□ strategy 159, 203
□ strategy for ～ 130
□ Streptococcus pyogenes 213
□ stress 168, 202
□ stress-related 175
□ striking 143
□ stroke 37
□ strong 128
□ structural 62, 102
□ structurally 51
□ structure 43, 65, 83
□ study 24, 154
□ study on ～ 140
□ study the role of ～ 108
□ subcutaneously 235
□ subject 83, 188, 204
□ submucosa 241
□ Subsequently, 103
□ subset 184
□ substrate 102, 120
□ substrate for ～ 176
□ subtractive 232
□ subtractive hybridization 232
□ subtype 163, 233
□ subunit 66
□ successful 238
□ such as ～ 61
□ sudden 146
□ suffer 36
□ suffer from ～ 36, 179
□ sufficient 149
□ sufficient for ～ 108
□ sufficient to ～ 199
□ suggest 23, 124
□ suggest a role for ～ 31
□ suggestion 145
□ suggest that ～ 52, 97, 101, 164, 214
□ sun 196
□ supplemental 202
□ support 77, 146, 150
□ support a model 31
□ support for ～ 134
□ suppression 225
□ suppressor 69
□ suprachiasmatic nucleus 46
□ suramin 92
□ surface 51
□ surface-exposed 26
□ Surprisingly, 101
□ surrogate 30
□ survival 36, 52, 64, 79, 110, 200, 232
□ survival rate 111
□ survivin 144
□ susceptibility 217
□ susceptibility to ～ 142, 172
□ susceptible 40
□ susceptible to ～ 40
□ suspect 64, 109
□ suspicion 190
□ switching 101
□ sympathetic 175
□ symptom 48, 122, 194, 202, 205, 238, 245
□ synapse 115, 153, 214
□ synaptophysin 116
□ syndrome 42
□ synergistic 181
□ synergy 150
□ synthesis 102
□ synthesize 128
□ synthetic 28
□ syphilis 152

- system ——— 59, 132, 150, 154
- systematically ——— 27
- system for ～ ——— 132
- systemic ——— 91, 180

T

- tadpole ——— 115
- tail ——— 203
- take care of ～ ——— 202
- take into consideration ——— 160
- Taken together, ——— 174
- take place ——— 244
- target ——— 96, 206
- targeted ——— 177
- target for ～ ——— 129
- T cell ——— 40
- T-cell receptor ——— 141
- technique ——— 134, 224
- technology ——— 79, 173
- telomerase ——— 203
- temozolomide ——— 165
- test ——— 52, 109
- testing ——— 89
- testis ——— 198
- test the hypothesis that ～ ——— 31
- tetramer ——— 141
- tetraploid ——— 235
- thallium ——— 218
- than expected ——— 181
- the fact that ～ ——— 195
- the hypothesis that ～ ——— 114, 134
- the idea that ～ ——— 146
- the majority of ～ ——— 38
- then ——— 163
- the notion that ～ ——— 146
- therapeutic ——— 111
- therapeutic effect ——— 111
- therapeutic index ——— 189
- therapeutic intervention ——— 57, 86
- therapy ——— 152, 230, 236, 243
- therapy for ～ ——— 131
- thereby ——— 163
- therefore ——— 160
- the view that ～ ——— 147
- think of ～ ——— 238
- thorough ——— 177
- thought to ～ ——— 56, 105
- threonine ——— 74
- thrombosis ——— 203
- Thus, ——— 162
- thyroiditis ——— 146
- tibia ——— 65
- tissue ——— 34, 134, 205
- tissue engineering ——— 67
- to a lesser degree ——— 93
- to a lesser extent ——— 189
- tobacco ——— 41
- to date ——— 123
- together ——— 134
- Together, ——— 134
- together with ～ ——— 96
- tolerated ——— 152
- tool ——— 131
- tool for ～ ——— 131
- to our knowledge ——— 236
- topoisomerase ——— 229
- to this end ——— 232
- toxic ——— 92
- trabecular bone ——— 65
- trace element ——— 106
- traditional ——— 196
- trafficking ——— 133
- trait ——— 24, 217
- transcript ——— 140
- transcription ——— 226
- transcriptional ——— 195
- transcriptional regulation ——— 57
- transdifferentiation ——— 154
- transfect ——— 76
- transfection ——— 42, 195
- transgene ——— 230
- transgenic ——— 85
- transient ——— 141, 195
- transiently ——— 78
- transiently transfected ——— 78
- translation ——— 156
- translocation ——— 80
- transmembrane segment ——— 127

□ transmit ―― 91
□ transplant ―― 235
□ transplantation ―― 161
□ transport ―― 85
□ transporter ―― 29
□ transposable ―― 230
□ transposable element ―― 230
□ transposition ―― 39
□ treat ―― 57, 92, 210, 242
□ treated with～ ―― 52
□ treatment ―― 86, 139, 158, 165, 169, 192, 214
□ treatment with～ ―― 143
□ trial ―― 80, 114, 165
□ trigger ―― 172
□ tuberculosis ―― 135
□ tubular cell ―― 235
□ tumor ―― 52, 59, 63, 112, 121, 206, 216
□ tumor cell ―― 51, 124, 135
□ tumor growth ―― 108
□ tumorigenesis ―― 229
□ tumor necrosis factor ―― 113
□ turnover ―― 176
□ typically ―― 171
□ tyrosine ―― 225
□ tyrosine kinase ―― 94

U

□ ultrasonographic ―― 27
□ ultraviolet ―― 196
□ unattainable ―― 158
□ unchanged ―― 136
□ unclear ―― 154
□ under conditions of～ ―― 228
□ undergo ―― 78, 171
□ underlie ―― 140, 229
□ underlying ―― 238
□ underlying mechanism ―― 46, 148
□ underscore ―― 199
□ understand ―― 131, 210, 226
□ understanding ―― 177
□ understand the mechanism ―― 32
□ undertake ―― 58
□ undertaken to～ ―― 58
□ under the control of～ ―― 230
□ under the same conditions ―― 228
□ unidentified ―― 97, 217
□ unilateral ―― 197
□ unique ―― 43
□ unknown ―― 68, 143, 169
□ unless ―― 216
□ Unlike ―― 159
□ unresolved ―― 74
□ unusually ―― 26
□ upstream stimulatory factor ―― 78
□ uptake ―― 42, 133, 136
□ urge ―― 202
□ urine ―― 95, 218
□ use ―― 76, 161
□ used as～ ―― 131, 204, 209
□ used by～ ―― 32
□ used for～ ―― 75
□ used to～ ―― 57
□ useful ―― 159
□ useful for～ ―― 108
□ useful in～ing ―― 24
□ use in～ ―― 243
□ use of～ ―― 205
□ usually ―― 35, 226
□ utility ―― 227
□ utilize ―― 150

V

□ vaccination ―― 224
□ vaccine ―― 61, 75, 114, 137, 151, 172, 238
□ vaccinia virus ―― 107
□ vaginal ―― 28
□ valuable ―― 23
□ value ―― 156
□ valve ―― 229
□ variant ―― 137
□ variation ―― 207
□ varied ―― 27
□ various ―― 150, 205, 244

□ vary 230
□ vascular 163, 180
□ vascular disease 32
□ vascular resistance 91
□ vector 61, 99
□ vein 65
□ ventricle 51
□ verbal 149
□ versatile 85
□ vertebrate 81, 237
□ very 152
□ vessel 65
□ viability 116
□ victim 241
□ view 47
□ viewed as ～ 47
□ viral 61, 161, 207
□ virulence 30
□ virulent 32
□ virus 49, 82, 91, 135, 137, 215
□ visual 114, 132, 204
□ vitamin 56
□ volunteer 151
□ vomiting 48

W

□ way 238
□ wealth of ～ 204
□ we attempt to ～ 44
□ we hypothesize that ～ 38
□ weight gain 188
□ well 152
□ well-defined 78
□ we propose that ～ 38, 225
□ we report that ～ 38
□ we speculate that ～ 39
□ whereas 155
□ while 154
□ white 186
□ white matter 51
□ widely 59, 157
□ widely used 75
□ wild-type 66, 111, 187, 193, 212, 215
□ Williams syndrome 80
□ withdrawal symptoms 154
□ within the context of ～ 229
□ with regard to ～ 243
□ with respect to ～ 124
□ word 194
□ work 169
□ writing 217

X

□ Xenopus laevis 115
□ xenotransplantation 41

Y

□ young 193
□ younger 236

Z

□ zebrafish 208
□ zinc 106, 128
□ Zucker rat 102

■ 著者略歴

河本　健
（かわもと・たけし）
広島大学大学院医歯薬学総合研究科助教．広島大学歯学部卒業，大阪大学大学院医学研究科博士課程修了，医学博士．高知医科大学助手，広島大学助手，講師などを経て現職．専門は，口腔生化学・分子生物学．概日時計の分子機構，間葉系幹細胞の再生医療への応用などを研究している．

大武　博
（おおたけ・ひろし）
福井県立大学学術教養センター教授．福井大学教育学部卒業，国立福井工業高等専門学校助教授，福井県立大学助教授，京都府立医科大学（第一外国語教室）教授などを経て現職．コーパス言語学の研究成果を英語教育に援用することが，近年の研究テーマである．

ライフサイエンス文例（ぶんれい）で身（み）につける英単語（えいたんご）・熟語（じゅくご）

2009年7月15日　第1刷発行
2018年3月20日　第5刷発行

著　者　河本（かわもと）　健（たけし）　大武（おおたけ）　博（ひろし）
監　修　ライフサイエンス辞書プロジェクト
発行人　一戸裕子
発行所　株式会社　羊　土　社
〒101-0052
東京都千代田区神田小川町2-5-1
TEL　03(5282)1211
FAX　03(5282)1212
E-mail　eigyo@yodosha.co.jp
URL　www.yodosha.co.jp/
印刷所　株式会社 加藤文明社

Printed in Japan
ISBN978-4-7581-0837-9

ライフサイエンス辞書プロジェクトの英語辞典

ライフサイエンス 論文作成のための英文法

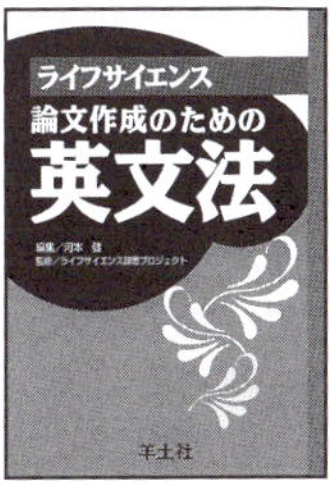

編集／河本　健
監修／ライフサイエンス辞書プロジェクト

論文でよく使われる文法・重要表現が一目でわかる！
論文執筆の基礎固めに最適の一冊！

定価（本体 3,800円＋税）　B6判　294頁　ISBN 978-4-7581-0836-2

ライフサイエンス 英語表現 使い分け辞典 第2版

編集／河本　健，大武　博
監修／ライフサイエンス辞書プロジェクト

論文ならではのコロケーションが数字でみえる！頻出の約1400語収載し，名詞には冠詞情報も追加された決定版！

定価（本体 6,900円＋税）　B6判　1,215頁　ISBN 978-4-7581-0847-8

ライフサイエンス英語 類語 使い分け辞典

編集／河本　健
監修／ライフサイエンス辞書プロジェクト

ネイティブならこう言い換える！約15万件の論文データから抽出された単語・表現でジャパニーズイングリッシュから脱出！

定価（本体 4,800円＋税）　B6判　510頁　ISBN 978-4-7581-0801-0

ライフサイエンス論文を書くための 英作文＆用例500

著作／河本　健，大武　博
監修／ライフサイエンス辞書プロジェクト

スラスラ書くコツは主語と動詞の選び方にあった！
とにかくすぐに書き始めたい人にオススメ！

定価（本体 3,800円＋税）　B5判　229頁　ISBN 978-4-7581-0838-6

発行 羊土社　ご注文は最寄りの書店，または小社営業部まで